THIS BOOK IS DEDICATED TO EVERYONE WORKING TOWARDS
AN ENVIRONMENTALLY SUSTAINABLE, SPIRITUALLY FULLFILLING
AND SOCIALLY JUST HUMAN PRESENCE ON THIS PLANET,
WITH DEEP GRATITUDE FOR YOUR COMMITMENT AND INSPIRATION.

ACKNOWLEDGMENTS

This Action Guide was created by an amazing team of concerned citizens – volunteers who also served as staff or board members – led by Maureen Jack-LaCroix of Be The Change Earth Alliance, a non-profit society based in Vancouver, British Columbia, Canada. Their commitment to take the journey of personal change by making sustainable, meaningful and just lifestyle choices has inspired thousands of others to co-create a better world, one action step at a time.

Maureen co-founded Be The Change Earth Alliance after completing a Masters in Eco-Psychology and Creation Spirituality through Naropa University, studying with Matthew Fox, Joanna Macy, Brian Swimme, Miriam McGillis, Drew Dillinger and other evolutionaries.

Bringing all the zeal of her former career as a special event producer to the priority mission of contributing to the global mind shift, Jack-LaCroix facilitated an eclectic team of professionals to focus their collective wisdom of personal development and behavioural change towards the planetary crises of today.

Executive coach, environmental scientist, spiritual director, social activist, behavioural therapist, addictions counselor, workshop leaders, facilitators, educators ... together they focused on bringing many successful processes for behavioural change into the context of helping people get personally involved in the ecological, social and spiritual crises of our times.

Special thanks to Suzanne Barois, Tyee Bridge, Susan Chambers, Erin Leckie, Jackie Larkin, Julia Morlacci, Ross Moster, Sharon Pendlington, Lucia Plescia, Kate Sutherland, Shane Wilson and Saskia Wolsak for their writing, research and editing contributions; and to Gregory Almas, Tracey Apple, Larry Butler, Marcus Hynes, Tosh Hyodo, Loretta James, Derek LaCroix, Carmen Mills, Toni Peroni, Bruce Sanguin, and Arunima Sharma for their inspiration and ongoing commitment to this Great Work.

Be The Change Earth Alliance bows deeply to wisdom elders such as Joanna Macy, Thomas Berry, Margaret Wheatley, Christine Baldwin, Deepak Chopra and respected indigenous peoples who have urged us to 'circle up' in the Great Turning.

Art Design and Layout: Katie Maasik
Cover photos: Jim Denham, John M. Cropper, Marisa DiFillippo

# TABLE OF CONTENTS

## VALUE B: REDUCE MATERIAL CONSUMPTION　　46

## VALUE C: CONSERVE ENERGY & NATURAL RESOURCES　　62

## VALUE D: RECONNECT TO SELF, OTHERS & EARTH　　85

# VALUE E: SUPPORT A GREEN & JUST SOCIETY   107

INTENTIONS

# VALUE F: GREEN MY COMMUNITY   124

INTENTIONS

# BIBLIOGRAPHY   140

# WELCOME TO BE THE CHANGE

## FOREWORD by BE THE CHANGE WISDOM CIRCLE

BE THE CHANGE is part of a huge social movement to create a new future for humanity. It's a significant undertaking that calls to the heart and soul of millions of people alive at this time—all those who can't stand by as their planet is despoiled and the lives of future generations are compromised.

This is our time to take a stand, by bringing the full awareness of our actions into every step we take. It's also our time to discover lives of greater fulfillment and joy. Nothing nourishes the soul like being the change you want to see in the world.

This Action Guide is your map to an exciting journey of personal choices, leading you to an empowered lifestyle and expanded world-view that will help sustain your health, your spirit, and the planet. It covers a full range of lifestyle choices and helps us become aware of how every single action we take either supports or undermines our vision for a sustainable, just and fulfilling future for all life.

May this Action Guide empower you to . . .

- *adopt the conscious practice of choosing actions that create the world you stand for*

- *strengthen your process of inner transformation, as you align thought, word, and deed*

- *connect to others on this path who can be partners in your journey to change the course of history*

No matter where you are in this continuum of change, we trust this guide will stimulate your imagination while offering new avenues of participation and unexamined assumptions to explore. It is also a great tool for change agents to educate and influence others.

It's more fun and more effective to do this with others-- so get together with neighbors, partner up with a friend! Trust us, we know you're busy. These days, we're all busy. But what is it worth to you to have a healthy home planet? Two hours every two weeks is a small commitment to make to changing the world, and most people are ready to make that commitment. We'll help you make the most of your time with a step-by-step guide to starting your own Action Circle. Give it a go!

Feel free to adapt the Action Guide and Action Circle program to your unique circumstances. It is designed to provide ample freedom for individual choice while offering enough structure to support and sustain you over time.

Every choice we make ripples into the world in mysterious and marvelous ways. Enjoy the experience of connecting with soul-mates in the Great Turning, and rest in Margaret Mead's wisdom that "a small group of thoughtful, committed citizens can change the world— indeed it is the only thing that ever has!"

# EDITOR'S / PUBLISHER'S PREFACE

CHANGE IS HAPPENING — everywhere, right now, exponentially! What an amazing time to be alive: a time of great responsibility and opportunity. With our civilization reaching a critical turning point, we are called to assess and shift our values and behaviours for the greatest good – not only for humanity's immediate future, but for all life on this planet.

In the face of severe ecological degradation, economic uncertainty and social inequality, at a time when traditional wisdom is trampled under mass consumerism, it is easy to despair that nothing we do can possibly make a difference. Think again! Absolutely everything we do makes a difference. This is what it means to "be the change you want to see in the world."

I have found it to be a profoundly humbling and inspiring practice to "be the change" by trying to shift my lifestyle and habits. As spiritual activist Mahatma Gandhi advised, "anything you do is insignificant, and everything you do is very important."

The simple actions in this Guide may indeed seem insignificant, but we have found that the process of change works on many levels. It's not only about doing the change, but also, indeed, about being the change. The practice of clarifying our true values and living them daily gives us the gift of knowing intimately both our responsibilities and our own power.

Einstein advised us that we can't solve the significant problems of today with the same level of thinking used when they were created. Once we start to make new and better choices in our lives, we inevitably influence our workplaces and communities.

In the field of human consciousness, everything we do has a ripple effect, and we can change cultural attitudes, awareness and beliefs in ways that are difficult to perceive but very real. Joanna Macy speaks of the Tibetan prophecy of the Shambhala warriors who would arise to help the world out of a dark time of destruction with two weapons: compassion and insight. So let's do what we can to be part of this global movement to restore the natural balance of life with heightened awareness.

Part of our evolving consciousness is to transform our firmly engrained "I" perspective into a "we." The 'circle' process outlined in this book is an ancient wisdom recommended by today's behavioral psychologists. When ancient wisdom and modern insight converge, the possibilities are very powerful.

Thank you for your willingness to start with this Guide and see where it leads you. It's an honour and privilege to join with you and support the Great Turning to a just, sustainable and spiritually fulfilled world.

Maureen Jack-LaCroix
Co-Founder, Be The Change Earth Alliance
Editor & Publisher, The Action Guide

# ABOUT THE ACTION GUIDE

The Action Guide is designed to help you make meaningful sustainability changes in your life: at home, at work and at play. It is a comprehensive road map that can grow to become a personal companion through your transformative experience.

There are so many aspects of our lives that are calling for re-alignment that at first it can seem a little daunting. Since everyone's situation is different, there are many options offered in these pages. Please note that the objective is not to do every action, but to find the ones that best enable us to affect positive change at this stage of our life.

At one time, we will be drawn to certain values and intentions... and later another value will call to us. Let this be an organic process as well as a practice of commitment. We try not to take on too much at once, and at the same time we strengthen our resolve to act upon what we believe in and bring our life energy to support our values.

## VALUES, INTENTIONS & ACTIONS

### 1st Tier: Values

If we are clear on our values, it is easier for our intentions and actions to naturally flow in alignment with who we are and how we want to be in the world. We offer a value statement for each section. You may want to write your own value statement to clearly articulate why this value is important to you. The material is organized into these value sections:

A – INCREASE HEALTH & WELL-BEING
*I value vibrant health for my body and mind, my children and family, my community and planet.*

B – REDUCE MATERIAL CONSUMPTION
*I value longevity, quality over quantity, sharing and spacious simplicity.*

C – CONSERVE ENERGY & NATURAL RESOURCES
*I value the limited gifts and life forms of the natural world and will not waste them thoughtlessly.*

D – RE-CONNECT TO SELF, OTHERS, AND EARTH
*I value feeling connected to myself and to others, both in human and natural communities.*

E – SUPPORT A GREEN & JUST SOCIETY
*I value my right to influence society and I support a socially just human presence on this planet.*

F – GREEN MY COMMUNITY
*I value resilient and healthy community and I will help create that for myself and others.*

### 2nd Tier: Intentions

Intentions represent a desire to make a change. Setting a clear intention can steer us in the right direction for bringing our lifestyles into alignment with our values. Each value has a variety of good intentions to choose from. For example, under VALUE A – Conserve Energy and Natural Resources, you will find INTENTION 1 – Drive Less.

### 3rd Tier: Actions

Once we are clear on our values and our intention, we are ready to get very specific about pledging an action we can actually achieve. This is a crucial step, because so often we can have good intentions that just never get acted upon. Every intention has a variety of actions to choose from – find the one(s) that appeal to you. These actions will help you to define manageable steps to make actual changes. For example, under VALUE C: *Conserve Energy and Natural Resources,* INTENTION 1 – Drive Less, you may choose the ACTION *"I will walk or bike to closer destinations ___ times this week."*

Research has shown that SMART goals (Specific, Measurable, Attainable, Realistic and Timely) are much more likely to be fulfilled, so you will find that the actions are also SMART. We encourage you to fill in the blanks and make the action your personal goal.

The actions are also tiered to reflect the various levels of commitment you may have for different intentions. We recognize that everyone enters this river of change through different streams, and while we may be very aligned in some values, we still have lots to learn in others.

## ACTION TIERS:

### The Basics
It's up to you to define how many times you will perform these actions. Taking the initiative even once can overcome a deeply engrained pattern and open the door to repeat behaviours and setting new habits. These first steps create a foundation to build on.

### I Can Do More
Beyond the basics, there is so much more we can do. Once we have set our intention, it's interesting to explore the many ways we can fulfill it. If you like to learn and grow, you'll have lots of fun here.

### Educating and Influencing Others
Once we've started changing our personal actions, it is natural to want to tell others about what we're up to, and passing on the word is one of our key roles as change agents. It's a good idea to hone our communication skills when we enter this phase, as people generally don't like being told what to do, and unbridled enthusiasm can provoke defensive reactions in others. We've put together a few communication tips on page 11 & 12 that we've found to be helpful. One of the easiest ways to influence another is to ask a friend to join you in one of your actions – this provides you with great support, something interesting to talk about, and simultaneously brings another good person onside.

### Community Projects
These actions are often more systemic initiatives that can be championed by an individual, but are easier to take on as a team. Reweaving healthy community is so important, and we need to look for every opportunity to turn our "I" into a "We." Use your creative imagination to consider how many of the other actions could also become community projects!

### Explorations 🌐
Learning more about these topics is easy thanks to the internet and its culture of information sharing and accessibility. Many of references in this Action Guide are internationally relevant and many are local to Vancouver Canada, because that's where the originators of this Action Guide live. Use their examples to find parallel resources that are relevant for your community.

### Measurements ✏
It is astonishing to calculate the cumulative impact of these simple actions if we continue to integrate this behaviour change throughout the year. Anytime you see a {pencil} you can visit www.bethechangeearthalliance.org/actionguide/measurables to obtain a simple spreadsheet that will provide quantitative feedback on how your actions are affecting the world.

# CHOOSING YOUR ACTION

### Trust Your Inner Guidance
Choose the action that calls to you – it will lead you forward. All the actions are inter-related and contribute to the process of shifting our overall perspective, so every choice is a good one.

### Customise Your Action
Feel free to modify any action to accurately reflect your commitment. Write it in your words. Make it specific and measurable. It doesn't matter if you make it more or less ambitious, just be sure to make it yours. If there's an action you want to take that isn't listed, then simply add it in!

You may also identify a new intention. This often arises when a circle decides to take on a collective project that will require a number of discrete actions to accomplish its goal. Start your own intention page under the appropriate value, and create a detailed action list as you fulfill your own intention.

If you would like to share your new intentions and actions to inspire others, please email us your additions and we'll include them on our website or in the next edition of this Action Guide.

### Maximize Your Impact
If you want to contribute the most positive environmental impact for your actions, focus on the intentions that are marked with an *. Actions within these intentions fall within the three categories of household spending that cause the greatest environmental impact, as identified by the Union of Concerned Scientists.[1]

# SUPPORT & INSPIRE

### Do It Together
Find a friend or family member to take on an action with you. Gather a few more friends and create an Action Circle! Your Action Circle may decide that it is most effective for pairs, small groups or the whole circle to choose the same intention for the week. You can also choose one intention coupled with a sustainability outing, like visiting a recycling depot. You create your own journey with the Action Guide!

### Share Your Stories of Change
We'd love you to share your story with the wider community by posting it on **www.mystandgo.com**. Stories are the basis of belief systems, and by sharing our personal experiences, we can inspire others to make similar shifts in their lives. As Einstein reflected, *"Often the things that count the most are not countable,"* and many of our actions fall into this domain. We always appreciate your insight and contributions.

### Record Your Pledges and Your Actions
Another great way to inspire and challenge ourselves and others is to pledge our intentions to others and then report on the actions we completed. This is one of the great dynamics of Action Circles. Every page of the Action Guide invites you to record your pledge date and the number of times you completed the action. The Action Guide also encourages you to continue your actions for 7 weeks so that they become firmly rooted as new 'good habits' in your life. You may also find it helpful to keep a dated record of the actions you pledged at the back of your Action Guide so that it is easy to reference.

---

1. A 1999 study by the Union of Concerned Scientists indicated that transportation, food and household operations/housing contributed 69% - 90% of the average U.S. household's impact on the environment. A list of 11 priority actions for U.S. consumers also developed from this study. The results of the study were published as a handbook "The Consumer's Guide to Effective Environmental Choices: Practical Advise from the Union of Concerned Scientists."

With the ease of social media keeping us connected on-line, we are now able to invite friends to join us and create Action Circles online. We even have a digital media application called AGENTS OF CHANGE that enable us to track our actions and challenge our friends to join us as an agent of change! Look for the Agents of Change link at www.bethechangeearthalliance.org to begin using this fun application to track your real world actions.

# EDUCATING & INFLUENCING OTHERS

After completing our own actions and changing our lifestyle, we can't help but want to educate and influence others. This is a worthy undertaking, and in order to be effective we would do well to examine some simple communication techniques. Communicating with others has a huge impact on our lives, and is too often taken for granted. Have you ever had the experience of being in the same conversation with someone, discussing the same topic, and finding out later that they left the conversation with a very different understanding of what transpired? This is a common experience! Successful companies often provide workshops to their management in order to decrease the costs of miscommunication.

Depending on how we communicate, our conversations can be compelling or repelling. We can create a sense of curiosity, empowerment and even inspiration to learn more or take action. Conversely, if we are insensitive and judgemental, we can offend and create resentment or disempowerment. Practicing good communication is an essential aspect of becoming an effective and approachable agent of change.

Here are a few key things to remember when speaking with others about sustainable lifestyle choices:

### Be Curious – Listen for Understanding
A conversation is two-sided. Addressing our values and behaviours can be interesting! If we truly listen to others, we are more likely to have an engaging conversation where we may both learn from one another. People react better to being heard versus being talked to. Try reflecting back what you heard, so that the others really *get* that you are listening. This also gives you an opportunity to adjust your message to suit what they are passionate about or can relate to. Be genuinely curious about their perspectives.

### Speak from personal experience
It is much easier to hear someone's point of view when it is shared as their own personal experience. Elaborate on why the topic at hand is a concern for you and what you are doing differently to help solve the problem. Referring to your own personal experience can open the conversation to asking about their experience. Do they have other ideas that you can consider? Being able to reference things first-hand has greater impact and is more empowering than simply reciting daunting facts. Using terms like "I found," "This makes me feel like," or "I like doing..." makes your communication more personal, may inspire curiosity, and is less imposing on others. Avoid phrases that start with "you" and especially "you should."

### Let go of judgement. Be respectful.
Everyone comes from different backgrounds, bringing different life experiences and diverse perspectives. Imposing judgement on another's process, understanding and actions makes potential collaboration and learning from each other far more difficult. When someone feels judged they are more likely to get defensive and not want to hear what you have to say. With humility and respect, we are able to learn from each other and relate at deeper levels. You will likely be surprised at what you are able to learn from others when you are able to listen without judgement.

### Your conversation is an offering - don't be attached to the outcome

Even though you may be very passionate about what you are talking about and willing to do anything for your cause, try to be sensitive to how overwhelming this energy can be to others. Unfortunately, people can easily feel guilty and close down defensively. Consider your message, and any educational information you are providing, as an offering. We suggest you practice being unattached to the outcome. If you impose expectation, then they are more likely to pull away from you and avoid future conversations. If you are calm and leave them with freedom to make their own choices in life, you will be feel empowered to sustain your role as a peaceful messenger.

### Be patient, yet persistent

Many people won't actually hear you the first time – this is not uncommon. It is said that it often takes seeing or hearing about something seven times before a person fully notices it and takes action. Be patient and continue supporting your cause and you will begin to notice changes taking place around you (You may discover that a particular conversation you had – weeks or months prior – was very meaningful to the other person and compelled them to change their habits!).

### Be as concise and clear as possible

This may take practice. Narrow down your message and clarify it to yourself before setting out to educate or influence others. Ask a question or two to set up your conversation. Reply with a few clear and concise sentences about why you care. This can be much more effective than reciting ideas that have been loosely pieced together.

### Use audience-specific language

When speaking to others, keep in mind where they are coming from and what type of language would be appropriate. If you are speaking to youth, use examples that they would relate to and terminology that they would understand. If you aren't talking to someone with a scientific background, then avoid using a lot of scientific words and references. Talk to them as a peer or as another potentially concerned citizen that may be taking some form of action in their life already. Never talk down to or belittle someone.

### Use appropriate body language

This is a fine art. When conversing with another, we are communicating with far more than just our words – our posture, distance from others, eye contact, voice fluctuations, pace, energy and enthusiasm all affect the message that is being relayed. Standing really close to someone can be intimidating, whereas avoiding eye contact and shying away can cause them to feel disengaged. Try to find a happy medium that others are comfortable with. If you are passionate and excited about your topic, you are more likely to pass that on and get others excited about it too (again, in moderation).

*"It is possible that the next Buddha will not take the form of an individual. The next Buddha may take the form of a community."* – Tich Nhat Hanh

# INTRODUCING ACTION CIRCLES

Countless wisdom elders, including Joanna Macy, Thomas Berry, Margaret Wheatley, Matthew Fox and Deepak Chopra have recommended that we undertake the challenge of this global shift in consciousness and behaviour by getting together in small groups or circles. It's an age-old methodology, and it works. Circles are the building blocks to the creation of a larger community of practice. This Action Guide can be used independently, but we highly recommend it be used to galvanize your own Action Circle. An Action Circle can be started by anyone. It can start with 2 or 3 people and build to the optimal 5-8 members who are friends, neighbours or co-workers that care about what's going on in the world and want to make a positive change.

Circles meet regularly, (weekly or bi-weekly works best) in homes, cafés, or wherever is convenient. As a circle member, we commit to 1 cycle at a time. One cycle is approximately 8 sessions long. Each session has a 1-2 hour structure that guides the circle process and still offers flexibility. It follows these basic steps:

- Welcome / Opening
- Purpose Statement
- Check-In
- Report on Actions Taken
- Dialogue
- Pledge New Actions
- Close

## KINDS OF ACTION CIRCLES

Action Circles can be adapted to suit our personal needs simply by adjusting the dialogue component. One element that carries us from session to session is the Action Guide. In addition to our dialogue focus, the Action Guide helps us create a personal practice of change. At every Action Circle, we choose an intention and pledge to take some related actions. It doesn't matter how many actions we take, only that they are meaningful to us. Our motto is "Act Big, Act Small, but Act!" Here are some examples of different kinds of Action Circles:

### Lifestyle Choices
Do you want to learn more about making sustainable lifestyle choices? You can use each value section of the Action Guide as the focus for your pre-reading and dialogue. There are lots of explorations to help satisfy your desire for more information on each topic. You can leave the dialogue component open for sharing what you learned from your explorations and actions taken.

### Community Project
Would you like to engage your neighbours on a community project? Use the dialogue component to identify the tasks that need to be accomplished in order to meet your objectives.

### Study A Sustainability Topic
If your circle wants to read and delve into a specific sustainability topic, there's an amazing selection of resources from Northwest Earth Institute, The Global Oneness Project and other organizations at www.bethechangeearthalliance.org on the Circle Convenors page.

### Advocacy Issue

Do you have a burning issue that you want to address? Bring together others who would like to focus on this topic with you. Set your goals and use the Action Circle format to sustain energy and organize your activities.

### Explore Topics

If there are a variety of interests within the circle, circle members can take turns preparing a 10-minute talk to focus the dialogue on the topic of their choice.

## THE POWER OF CIRCLES

The illusion of separation is a huge obstacle to sustainable living, so the act of bringing people together is a vital part of the solution. Action Circles provide the lively, community-based support we all need to shed old habits, make new lifestyle choices, and spark significant social change. Research and testing has shown that Circles can help us:

- Deepen our understanding of the key issues facing humanity through meaningful dialogue
- Make sustainable lifestyle choices that align to our values
- Provide each other with peer support and accountability for great results,
- Build community for group projects and collective actions

Action Circles are designed to engage the heart, the head and the hand, supporting us to use compassion, wisdom and action in order to affect positive change in the world. We also know that it can be hard to do this work alone, which is why peer support and accountability empower the Action Circle along the way.

### Heart

Meeting in Circles helps us to reweave community connection. Circles engage our hearts by creating cultures of mutual support and care. What can seem too difficult to tackle on our own becomes quite manageable with the help of our peers. Reciprocally, we expand our motivation and focus from self to others.

### Head

To help deepen understanding and engage our minds, we offer a variety of reading, video, and online resources from collaborating organizations. Learning about our global situation is greatly deepened in dialogue. We voice our own perspectives and we expand our abilities to understand the world as we become enriched by the collective wisdom of the entire circle.

### Hand

It is very satisfying to ground our informed concerns by taking action. At each circle gathering, we report back on the actions we chose from the week before, share our stories of challenge and victory, and then choose the next action. The cumulative efforts of the entire circle can inspire us to see how small actions can add up to effect major change.

*"I can't change the world on my own, I need at least three people to do it."*

– Bill Mollison, co-originator of Permaculture

### Peer Support

One of the greatest gifts the circle format offers is mutual support for following through on one's best intentions. Research shows that what we focus on increases, so we appreciate all successes, big and small! There are a few simple guidelines that are encouraged in Action Circles:

- Circle mates are acknowledged for every change.
- They are also encouraged to be curious about the shift in values that their changes may reveal.
- If someone is unsuccessful in fulfilling the action they pledged, they can choose to try again next week. Circle mates are encouraged to help others with their actions. A mid-week reminder email or sharing some of the research can help a great deal!

### Accountability

Accountability in an Action Circle greatly enhances a person's probability of success in affecting personal change. A business coach who worked with us viewed the circle as a "specific accountability appointment." She offered us the following statistics on the probability of accomplishing our goals:

| If you: | Your Success Rate Is: |
|---|---|
| Come up with the idea to attain a goal... | 10% |
| Consciously decide to attain this goal... | 20% |
| Set a date to have the goal completed by... | 40% |
| Put the goal in place... | 50% |
| Commit your goal to someone else... | 65% |
| Have a specific accountability time for the completion of your goal... | 95% |
| *This is what happens in Action Circles!* | |

Source: American Society for Training and Development

Action Circles make it easy to achieve our goals. This combination of heart, head, hand, peer support and accountability is a tested model; we quickly experience the connection between our values, our lifestyle choices, our commitment to change, and the health of our planet.

## THE ACTION CIRCLE PROCESS

The true power of any Action Circle lies in the collective wisdom of the group. Action Circles are a co-creation, meaning that no one individual person is in charge. There are, however, a few roles that can help ensure that your Action Circles run smoothly.

### Convenor/Host

As a host, you act as a nucleus, attracting other concerned people into your aura of influence and offering ways to focus concerns in productive ways. You gather circle members, call and organize the circle location, take care of administrative duties, and keep the group connected and informed.

### Facilitator

The facilitator's job is to direct the circle conversation. This simply means knowing the agenda for each circle and ensuring that there is enough time to get all the items complete by the end of each session. Circles may choose to rotate facilitators so that everyone has a turn leading the process.

Experienced facilitators also speak of "holding the group energy" – a kind of attention that helps everyone feel welcome and cared for. If you are looking for more facilitation guidance and a wide variety of group activities for facilitators, check out the Circle Guide at:
**www.bethechangeearthalliance.org**

### Cycles

Action Circles are organized into a cycle of approximately 8 sessions. We have found that completing a cycle allows everyone to mark their progress. This also creates space for new members to join and for others to take time out. Cycles help keep circle energy fresh and alive. Your first and last Action Circle will be a little different than your ongoing Action Circles.

### Tracking Your Actions

It is rewarding to track your process and it encourages others to see what you have accomplished. Join the Agents of Change social media application that lets you track your real-world actions for points that clean up your virtual world!

## ACTION CIRCLE AGENDAS

This basic agenda creates a simple container to hold the energy of your Action Circle:

1.    WELCOME/ OPENING
2.    PURPOSE STATEMENT
3.    CHECK-IN
4.    REPORT ON ACTIONS TAKEN
5.    DIALOGUE
6.    PLEDGE NEW ACTIONS
7.    HOUSEKEEPING
8.    CLOSE

The following outline for your Action Circle is designed to be flexible while providing enough structure to feel complete with a start, middle, and end. You are free to adjust the format as you like, as long as you cover all of the bases. For example, some circles prefer to report on their actions with their check-ins and make their pledge for next actions at the same time.

The timing is also flexible, so feel free to make the circle as long or as short as everyone is comfortable with. For example, some circles prefer to have a 'soft start' with some social tea time that precedes the opening.

*"If success or failure of this planet and of human beings depended on how I am and what I do, How Would I Be? What Would I Do?"* – R. Buckminster Fuller

## Basic Action Circle Agenda (1.5 hours)

| | | |
|---|---|---|
| OPENING & PURPOSE STATEMENT | 2 MIN. | Facilitator takes a moment to focus the group's energy and recite the purpose statement of the circle. |
| CHECK-INS | 15 MIN. | Each person checks-in. Where are you at today? For groups of 5 people, allow 2.5 minutes per person; for groups of 10, allow 1.5 minutes. |
| REPORT ON YOUR ACTIONS | 15 MIN. | Go around the circle and share the experiences you had, positive or negative, with the actions you pledged from your last session. |
| DIALOGUE | 40 MIN. | Focus on your sustainability topic (community garden, local food production, oil tankers, etc.). |
| PLEDGE NEW ACTIONS | 10 MIN. | Go around the circle and choose new actions (or stick with what you have!) to complete for the next session. |
| HOUSEKEEPING | 5 MIN. | Any details about who is facilitating next circle, or bringing refreshments, etc. |
| CLOSE | 3 MIN. | Go around the Circle; each person shares something that they are grateful for. The facilitator can offer final remarks. |

# FIRST ACTION CIRCLE

The first Action Circle of a cycle needs to allow for more time to complete these additional items;

## Choose Your Topic/Focus
Either the host chooses the focus before starting the circle, or circle members decide now.

## Hand out Action Guides
Either they are ordered in advance or circle members place their orders during the first session. Take some time to introduce the Action Guide and how it works.

## Establish Group Guidelines/Community Agreements:
Having had some time to feel the group dynamics, begin to list interaction guidelines that everyone agrees to on a piece of paper (arrive on time, 2-second pause between talking, confidentiality, etc.). Keep this document visible at each circle for easy reference by everyone.

# COMPLETION ACTION CIRCLE

## Action Report
In addition to reporting actions pledged at last circle, take time to reflect on all the actions that you completed and the shift in world view that you experienced over the past cycle.

## Dialogue Focus
Share what you enjoyed about the circle experience and any challenges you encountered. Decide if you are going to continue on with another topic, have new members join, or wrap up this circle for now.

## Celebration
This is a perfect time for a pot luck social. Great to invite friends who may want to join the next circle cycle.

# VALUE A
# INCREASE HEALTH & WELL-BEING

*I value vibrant health for my body and mind,*
*my children and family, my community and planet.*

Medical science and technology are performing miracles in the operating room, and for this we can be grateful. But environmental toxins, the stress of work-addicted lifestyles, and the altered food we consume are breaking down our bodies' immune systems and creating disease. Rates of cancer are rapidly escalating, along with many other issues like infertility and obesity[1]

Now more than ever before, we need to value ourselves and the world enough to take care of our bodies and minds. We create endless benefits for ourselves and others when we choose to make that investment of time, awareness and energy. Healthy food choices and regular exercise can save thousands of dollars in medical bills later in life or even next year—not to mention increasing our happiness, productivity, self-confidence and our ability to handle stress.[2] [3]

By taking action to limit our exposure to toxins and reduce stress, we are investing in our quality and longevity of life. Naturally enough, taking care of our own health and well being also creates health and well-being for the planet.

## I Intend To:

| | |
|---|---|
| A 1 | **Limit the Incoming** |
| A 2 | **Do Less, Enjoy More** |
| A 3 | **Take Control of the Electronics** |
| A 4 | **Get Outside & Get Fit** |
| *A 5 | **Eat Less Altered Food** |
| A 6 | **Cut Back on Easy Addictions** |
| A 7 | **Eat Ocean Wise** |
| A 8 | **Eat Happy Meat** |
| A 9 | **Eat Less Meat** |
| A 10 | **The Art of Composting** |
| *A 11 | **Toxin Free Gardening** |
| A 12 | **Say No to Toxins** |
| A 13 | **A Toxic Free Household** |
| A 14 | **Reduce Effects of Radiation** |

1. www.aacc.org/publications/cln/2009/july/Pages/newsbrief0709.aspx
2. www.happinessandwisdom.com/work-out-journal/exercise-endorphins-and-happiness
3. www.livestrong.com/article/274558-what-are-the-effects-of-eating-healthy-exercising/

# INTENTION 1 LIMIT THE INCOMING

With so many competing demands on our attention and time, we can easily feel overwhelmed by the flood of information we are receiving. While attempting to process and respond in a timely manner we can lose sight of our real priorities. Controlling the incoming of information and demands is an important way to create the space and time we need to reflect upon our direction and align our time allocation with our values. Giving our internal systems a chance to rest and rebalance is a crucial part of maintaining health and well-being.

| Actions | The Basics | PLEDGE DATE | # TIMES/ WEEK | # TIMES COMPLETE / 7 WEEKS |
|---|---|---|---|---|
| A.1.1 | I will minimize unnecessary items and clutter from my home. | | | |
| A.1.2 | I will minimize unnecessary items and clutter from my office. | | | |
| A.1.3 | I will limit the amount of incoming e-mail messages and schedule replying by trying the following practice: Example: Review and Reply to emails for one hour at the beginning and end of each working shift. | | | |
| A.1.4 | I will not use my cell phone for ___ days/hours this week. | | | |
| A.1.5 | I will limit my subscriptions to newspapers, magazines and newsletters by cancelling ____ subscriptions. | | | |

| Actions | I Can Do More | | | |
|---|---|---|---|---|
| A.1.6 | I will cancel ____ catalogue subscriptions and choose to browse products online instead. | | | |
| A.1.7 | I will organize my emails into "lists" for announcements from organizations that I'm interested in, but can't keep up with on a daily basis.  | | | |
| A.1.8 | I will "purge" ___ drawers or countertops this week.<br>1. Empty everything from the drawer or shelf into a pile.<br>2. From this pile, pick out only the most important things, the stuff I use and love.<br>3. Get rid of the rest. Right now. Recycle it, or give it away.<br>4. Put the stuff I love and use back, in a neat and orderly manner. | | | |
| A.1.9 | I will limit my memberships in clubs and organizations to those I am actively involved with. I will cancel ___ memberships. | | | |
| A.1.10 | I will limit the number of credit cards I use. I will cancel ___ credit cards. | | | |
| A.1.11 | Here's a specific way I can limit what's coming into my life:_____ _____ | | | |

## EXPLORATIONS

For lots of great ideas on simplifying your life:
*www.simpleliving.org*
*www.zenhabits.net*

# INTENTION 2  DO LESS, ENJOY MORE

How often have you found yourself having too many things to do, not enough time to do them, and wanting more hours in a day? If you answered, "Too often!", then we invite you to make a radical decision to simply do less!

Chronically overextending ourselves with too many activities and commitments ultimately takes a serious toll on our wellbeing at all levels—physically, emotionally, and intellectually, and decreases our ability to be truly effective in the world. Choose to make some of the following activities a priority and enjoy the benefits of a less harried lifestyle and a calmer inner life.

| Actions | The Basics | PLEDGE DATE | # TIMES/ WEEK | # TIMES COMPLETE / 7 WEEKS |
|---------|-----------|-------------|---------------|----------------------------|
| A.2.1 | I will be economical and discriminating with my time by thinking before I say "yes". | | | |
| A.2.2 | I will create "space" in my life by not scheduling activities back to back. | | | |
| A.2.3 | I will give full attention to the activities I am engaged in. | | | |
| A.2.4 | I will practice being "offline" for ___ day(s) this week. | | | |
| A.2.5 | Before accepting a new activity, I will end an old one. | | | |

| Actions | I Can Do More | | | |
|---------|--------------|---|---|---|
| A.2.6 | I will take a Time Inventory this week. | | | |
| A.2.7 | *How much time did I spend this week on productive work, on organizing my home and life, on renewing my energy and spirit, and on reflecting on personal goals & dreams?* I will journal the changes I would like to make. | | | |
| A.2.8 | I will scale back projects and enjoy the process. | | | |
| A.2.9 | I will not work more than 8 hours a day. | | | |
| A.2.10 | I will stop doing something that causes me to feel stressed and unhappy this week. I will stop doing: _____ | | | |
| A.2.11 | I will start doing something this week that brings me joy. | | | |
| A.2.12 | I will limit the number of evenings I am away from home to ___ nights. | | | |
| A.2.13 | I will make a "Most Important Tasks" (MITs) list each day, and setting no more than 3 very important things I want to accomplish each day. (A simple list of 3 things that make me feel like I've accomplished something.) 🌐 | | | |
| A.2.14 | I commit to doing action #(s) ____ consistently for ____ days/weeks/months. | | | |

# INTENTION 2 DO LESS, ENJOY MORE

| | | PLEDGE DATE | # TIMES/ WEEK | # TIMES COMPLETE / 7 WEEKS |
|---|---|---|---|---|
| ACTIONS | **Educating and Influencing Others** | | | |
| A.2.15 | I will ask a friend who seems stressed what they could do to simplify their life, cut back to only 8 hours a day, or simply to leave work at work. | | | |
| A.2.16 | I will commit my family to a no plans day once a month/week. We will make no plans for work or play and just be together and hang out on _____day. | | | |
| A.2.17 | I will ask a friend, coworker or family to join me in completing action #(s) _____ from this intention. | | | |

## EXPLORATIONS 🌐

Most Important Tasks (MITs) lists: *www.zenhabits.net*

Using your time wisely articles: *www.easytimetracking.net/article8.asp*

Read: *The Power of Less: The Fine Art of Limiting Yourself to the Essential in Business and Life.* by Leo Babauta, Hay House UK, 2009.

# INTENTION 3 TAKE CONTROL OF THE ELECTRONICS

With all our electronic technology enabling high speed communication, we are living in the "Age of Information." Although easy access to information and entertainment can be valuable, our electronic devices can also become addictive and huge time-wasters. If we are not mindful and disciplined about our habits, we can easily displace activities that are more valuable for our overall wellbeing and effectiveness. The average American spends a substantial portion of their life detached from day to day activities and attached to electronics. For example, the average American spends more than 4 hours watching television each day, or 28 hours a week – that's 2 months of non-stop TV watching per year.[1] In turn, that's 9 years of your life spent watching TV by the time you are 65 years old. It's valuable to notice when we are drawn to these addictive behaviours and how we feel when we decide not to do them. In the space that is created, we have valuable opportunity to choose something that brings us joy. What might that be?

| | | PLEDGE DATE | # TIMES/ WEEK | # TIMES COMPLETE / 7 WEEKS |
|---|---|---|---|---|
| ACTIONS | **The Basics** | | | |
| A.3.1 | I will count the number of hours I spend watching TV this week, and reduce this by ___ hours the following week. | | | |
| A.3.2 | I will track the number of hours I spend on the computer this week, and reduce this by ___ hours the following week. | | | |
| A.3.3 | I will list the activities I would like to do if I had more time and reward myself with these options. Example: exercise, drawing, music, dining with friends, visiting with family, reading. | | | |

1. www.csun.edu/science/health/docs/tv&health.html

# INTENTION 3  TAKE CONTROL OF THE ELECTRONICS

| | | PLEDGE DATE | # TIMES/ WEEK | # TIMES COMPLETE / 7 WEEKS |
|---|---|---|---|---|
| A.3.4 | I will educate myself about electromagnetic safety and take an action to reduce my exposure, in specific:_____. 🌐 | | | |
| A.3.5 | I commit to doing action #(s) ___ consistently for ___ days/weeks/months. | | | |

| ACTIONS | **I Can Do More** | | | |
|---|---|---|---|---|
| A.3.6 | I will track the number of hours I spend on Facebook or other electronic social connectivity sites this week, and reduce this by ___ hours the following week. | | | |
| A.3.7 | I will turn off the TV except for specific shows that I want to watch and not watch for more than 2 hours at a time. | | | |
| A.3.8 | I will schedule personally interactive activities in my day, to break up the amount of time I am on the computer. | | | |
| A.3.9 | I will track the amount of calls I receive and send on cell phone this week, and reduce this by ___ calls the following week. | | | |
| A.3.10 | Video games, movies, and other media are replacing my time to explore my own life. This week I choose to abstain from the following activity to explore the space this creates for me: _____ | | | |
| A.3.11 | I will ask a friend, coworker or family to join me in completing action #(s) _____ from this intention. | | | |

| ACTIONS | **Educating and Influencing Others** | | | |
|---|---|---|---|---|
| A.3.12 | I will get my family to reduce their TV time at home by ___ hours. | | | |
| A.3.13 | I will get my family to reduce their computer time by ___ hours. | | | |
| A.3.14 | Our household will have a non-electronics meal (TV, phones, MP3 players and computers). | | | |
| A.3.15 | I will ask a friend, coworker or family to join me in completing action #(s) _____ from this intention. | | | |

| ACTIONS | **Community Projects** | | | |
|---|---|---|---|---|
| A.3.16 | We will advocate an electronics-free day in our community, either annually or monthly. | | | |
| A.3.17 | We will personally implement action #(s) _____ from this intention at our schools and/or workplaces. | | | |
| A.3.18 | We will request that our workplaces and/or schools consider implementing action #(s) _____ from this intention; we will research how this is done to make the process as easy as possible. | | | |

# INTENTION 3  TAKE CONTROL OF THE ELECTRONICS

## EXPLORATIONS 🌍

Electromagnetic Safety: *http://www.powerwatch.org.uk/*
*www.citizensforsafetechnology.org/*
*www.icems.eu/public_education.htm*
Read: *Last Child in the Woods: Saving our children from nature deficit disorder.* by Richard Louv, Algonquin Books
of Chapel Hill, 2005.

## INTENTION 4 GET OUTSIDE AND GET FIT

Many of us spend our days in sedentary positions (sitting at a desk in front of a computer at work), only to come home and spend more time in sedentary activities. Yet our bodies require regular movement for our muscles, bones, lungs, heart, etc., to stay toned and healthy, reducing the chances of getting ailments such as diabetes, cancer and heart disease. Just 15-45 minutes of activity a day, several times a week, can help us burn off the harmful effects of stress, strengthens our ability to respond effectively to stressful situations, work as an anti-depressant, boost our confidence and looks, strength and flexibility.[2] [3] If you want to live longer, healthier and happier, commit to your body and get out there and be active.

| Actions | The Basics | PLEDGE DATE | # TIMES/ WEEK | # TIMES COMPLETE / 7 WEEKS |
|---------|-----------|---|---|---|
| A.4.1 | I will take a walk after dinner. | | | |
| A.4.2 | I will take a 2 hour or longer walk in nature. | | | |
| A.4.3 | I will attend a drop-in yoga or pilates class. | | | |
| A.4.4 | I will go for a bike ride or go rollerblading. | | | |
| A.4.5 | I commit to doing action #(s) ____ consistently for ____ days/weeks/months. | | | |

| Actions | I Can Do More | | | |
|---------|--------------|---|---|---|
| A.4.6 | I will start an exercise program that's good for my body (gym, yoga, pilates, walking, running). I will do this activity ____ per week for ____ weeks. | | | |
| A.4.7 | I will try out a recreational physical activity that I don't do on a regular basis, specifically: _____. | | | |
| A.4.8 | I will avoid engaging in an outdoor activity that has a negative impact on the environment. Example: Riding snowmobiles, off-road vehicles, ski-jets, and powerboats, golfing on chemically treated greens. | | | |

2. http://www.happinessandwisdom.com/work-out-journal/exercise-endorphins-and-happiness
3. http://www.livestrong.com/article/471729-how-are-endorphins-associated-with-exercise/

# INTENTION 4  GET OUTSIDE & GET FIT

| | | PLEDGE DATE | # TIMES / WEEK | # TIMES COMPLETE / 7 WEEKS |
|---|---|---|---|---|
| A.4.9 | I will support the local economy when traveling by staying in locally owned lodgings and traveling via local transport. | | | |
| A.4.10 | I commit to doing action #(s) ___ consistently for ___ days/weeks/months. | | | |
| Actions | **Educating and Influencing Others** | | | |
| A.4.11 | I will engage my family in outdoor activities involving personal interaction. | | | |
| A.4.12 | I will plan a vacation with others in a way that saves resources. Example: accommodations and transportation. | | | |
| A.4.13 | If I have small children in school, I will help get a program going that encourages the students to walk and bike. Example: "walking school bus" or Bike Smarts.  | | | |
| A.4.14 | If golfing, I will make inquiries and ask the golf club to use alternative measures to toxic chemical pesticides and herbicides often used on golf courses. 🌐 | | | |
| A.4.15 | I will ask a friend, coworker or family to join me in completing action #(s) _____ from this intention. | | | |
| Actions | **Community Projects** | | | |
| A.4.16 | We will request that our municipal planners and politicians create integrated transportation networks that provide space and facilities for cyclists. For example, bicycle paths, bike-and-ride centres, and bike lanes. | | | |
| A.4.17 | We will start a "Take a hike" club where groups of friends, family members, coworkers or neighbours meet weekly or biweekly to bike ride, snowshoe, paddle, or _____. | | | |
| A.4.18 | We will start an active lifestyle club, or introduce exercise clubs into our workplaces. | | | |

## EXPLORATIONS 🌐

Research the topic of green recreation and try something new: *www.greenlivingideas.com/topics/health-and-fitness/ recreation-and-outdoors/eco-friendly-outdoor-recreation*

Activities to keep our youth busy and happy in nature: *http://richardlouv.com/books/last-child/resource-guide/*

To help inspire your team members, sports federations, and fans about solutions to global warming, join Play It Cool. Visit: *http://www.davidsuzuki.org/issues/climate-change/science/play-it-cool/index.php*

Research golf courses and pesticide and herbicide issues and what is being done by golf courses for healthier grounds: *http://www.beyondpesticides.org/index.html*

When you travel, don't leave your values and social consciousness behind. Check out agencies like Tourism Concern, a resource for ethical and sustainable tourism: *http://www.tourismconcern.org.uk/*

Refer to Low Carbon Diet by David Gershon (2007), page 44, and learn about the Journey for the Planet school curriculum by visiting *www.empowermentinstitute.net/lcd*

# *INTENTION 5 EAT LESS ALTERED FOOD

There is a well known phrase "we are what we eat." Most of us aspire to be good, whole, vibrant and healthy people. But often we are feeding ourselves altered, compromised, over-processed or over-cooked food. And then we wonder why it is so hard to stay totally healthy, focused and vibrant. Many mainstream foods we eat are genetically modified, non-organic, altered to last longer and wrapped or heated in plastic. The majority of animals that are fed large portions of genetically modified foods have dramatically increased mortality and infertility rates, along with many other scary side effects.[4]

The function of pesticides and herbicides that are used in non-organic foods are to kill other plants, insects and animals; in turn this is what we are eating and ends up in our soils and water systems. Preservatives are carcinogenic and linked to neurological and behavioral issues.[5] Water that has been in contact with many plastics has been linked to both increased estrogen levels and causes various cancers and ecosystem issues. The facts speak loudly. It is time for us to take responsibility for our health in our everyday actions and in turn support the healing of our planet.

| Actions | The Basics | PLEDGE DATE | # TIMES/ WEEK | # TIMES COMPLETE / 7 WEEKS |
|---|---|---|---|---|
| A.5.1 | This week, I will avoid preservatives and other artificial ingredients in my food. I will not eat any processed food for ___ days or weeks. | | | |
| A.5.2 | I will buy organic vegetables and fruit. *On average, organic vegetables contain 40 percent more vitamins and antioxidants when compared to conventional, agri-business produced vegetables and lower levels of heavy metals, mycotoxins and pesticide residues.[6]* | | | |
| A.5.3 | I will buy organic dairy products. Example: Milk, cheese, yogurt, etc. | | | |
| A.5.4 | I will not heat any food in plastic. | | | |
| A.5.5 | I will not eat food cooked or heated in a microwave oven. | | | |
| A.5.6 | I will do research on Genetically Modified (GM) foods, which foods are GM and how they affect me. 🌍 | | | |
| A.5.7 | I will go to the grocery store and notice which non-organic products I buy have corn, wheat, soy, dairy or other commonly GM foods. *Approximately 70% of processed food in North American grocery stores contains genetically modified (GM) ingredients.[7]* | | | |

| Actions | I Can Do More | | | |
|---|---|---|---|---|
| A.5.8 | I will research what organic really means and what its health benefits are for me and my environment. | | | |
| A.5.9 | I will prepare ___ meals this week made with only organic ingredients. | | | |

4. http://livinglime.ca/2009/05/21/scary-gmo-studies/
5. http://ebrandaid.com/additives-preservatives/7-scary-things-lurking-in-your-food/
6. http://www.livestrong.com/article/333160-the-advantages-of-fresh-vegetables-picked-from-the-garden/#ixzz1Sa3zZUJZ
7. http://www.drheise.com/geneticfood.htm

# INTENTION 5  **EAT LESS ALTERED FOOD**

| | | PLEDGE DATE | # TIMES/ WEEK | # TIMES COMPLETE / 7 WEEKS |
|---|---|---|---|---|
| A.5.10 | I will not eat any GM food for ___ days or weeks. | | | |
| A.5.11 | I will do research on what our compulsory labeling requirements are, and those of other countries. | | | |
| A.5.12 | I will only buy food that I know what each ingredient is for ___ days or weeks. | | | |
| A.5.13 | I commit to doing action #(s) ___ consistently for ___ days/weeks/months. | | | |
| ACTIONS | **Educating and Influencing Others** | | | |
| A.5.14 | I will research food additives this week and post information in a common area. | | | |
| A.5.15 | I will hold an organic, non-GMO and preservative free pot-luck or dinner. | | | |
| A.5.16 | I will ask a friend, coworker or family to join me in completing action #(s) _____ from this intention. | | | |
| ACTIONS | **Community Projects** | | | |
| A.5.17 | We will suggest healthier food options at the cafeterias at our schools and/or workplaces. | | | |
| A.5.18 | We will research healthy cafeteria programs in our region and bring our findings to a decision-maker at our school and/or workplace. | | | |
| A.5.19 | We will encourage ___, advocate ___, and ensure ___ that we buy antibiotic- and hormone-free and/or organic cream and milk this week for organizational use. | | | |
| A.5.20 | We will find out who is currently lobbying for mandatory GM food labeling in our country and join their cause. | | | |
| A.5.21 | We will write letters advocating mandatory labeling. | | | |
| A.5.22 | We will source a sustainable company that offers fair-trade organic tea and/or coffee for our next organizational meeting, party or event. | | | |

## EXPLORATIONS

Research why it is important to eat organic foods and to limit the additives, antibiotics, and hormones currently in the food we eat. For information see the Planet Friendly synopsis at *www.planetfriendly.net/organic.html*
Genetically Modified (GM) Food Crop Information:
*http://www.npr.org/templates/story/story.php?storyId=129010499*
World Watch: *http://www.worldwatch.org/node/5950*
*www.ota.com/organic_and_you/10reasons.html*
Long term effects:
*http://www.infowars.com/gmo-scandal-the-long-term-effects-of-genetically-modified-food-on-humans/*
Make your community GM-free:
*http://www.organicconsumers.org/* or *http://www.canadians.org/food/index.html*

## INTENTION 5  EAT LESS ALTERED FOOD

Food: *The Ultimate Secret Exposed - PT 1/2*
*http://www.youtube.com/watch?v=MSpkLk0vYmk*
*The Omnivore's Dilemma.* by Michael Pollan, Penguin Press, 2006.
Become informed about the food you eat, and how you can eat more sustainably, by referring to online websites
and magazines, such as *http://www.sustainabletable.org/home.php*
*http://www.eatwellguide.org/i.php?pd=Home*
"Eating Matters" page of the Huffington Post: *http://www.huffingtonpost.com/news/politics-of-food*

# INTENTION 6 CUT BACK ON EASY ADDICTIONS

There are so many things in our life that we end up subconsciously relying on. When we become more aware of the little or big addictions in our lives and consciously choose to cut back or do without, we strengthen and empower our will for self love and healthy self care. Knowing that we don't need these ingredients in our lives, helps us choose what we do want to consume and supports a healthy metabolic balance. Snack foods, cereals and other prepared meals have been found to be able to trigger the brain in the same way as tobacco. Many food manufactures have combined fat, sugar and salt in a manner that is so appealing that it is hard for one to stop eating the product, even after you are full.[8] Choose to make conscious choices about how you feed and fuel your body, mind and emotions.

| Actions | The Basics | Pledge Date | # Times/ Week | # Times Complete / 7 Weeks |
|---|---|---|---|---|
| A.6.1 | I will reflect on my diet and write a list of the different substances I feel I may be addicted to or have a hard time reducing. I will contemplate how each substance does or doesn't serve me. | | | |
| A.6.2 | I will not add any sugar to my food or beverages for___ days or weeks. | | | |
| A.6.3 | I will not buy any drinks or products sweetened by high-fructose corn syrup (glucose-fructose syrup) for __ days or weeks. HFCS is linked to excessive weight gain, as well as other serious health conditions. 🌍 | | | |
| A.6.4 | I will not have any products with aspartame in it for __ days or weeks. Aspartame is addictive and has been linked to such neurological conditions such as epilepsy. 🌍 | | | |
| A.6.5 | I will refrain from buying salty snacks ____times. | | | |
| A.6.6 | I will limit my caffeinated beverages to __ per day or week. | | | |
| A.6.7 | I will monitor how many cigarettes or cigars I consume in a week and will reduce this number by ___ for the following week. | | | |
| A.6.8 | I will monitor how much alcohol I drink over the next week and reduce that by ___ drinks or oz. the following week. | | | |

8. telegraph.co.uk/foodanddrink/foodanddrinknews/5673512/Why-junk-food-really-is-addictive.html

# INTENTION 6  CUT BACK ON EASY ADDICTIONS

| | | PLEDGE DATE | # TIMES/ WEEK | # TIMES COMPLETE / 7 WEEKS |
|---|---|---|---|---|
| Actions | **I Can Do More** | | | |
| A.6.9 | I will go to the doctor and have my nutrient and blood levels checked out for high blood sugar, cholesterol, etc. | | | |
| A.6.10 | I will not have any soft drinks for ___ days or weeks. | | | |
| A.6.11 | I will not eat sweet desserts or snacks for ___ days or weeks. | | | |
| A.6.12 | I will not eat any foods with added sugar, aspartame or high-fructose corn syrup for ___ days or weeks. 🌐 | | | |
| A.6.13 | I will not go over my DV % (daily value) of sodium (salt) in the products I purchase for __ days or weeks. | | | |
| A.6.14 | I will not add any salt to my meal when I usually would. | | | |
| A.6.15 | I will not have a cigarette or cigar or __ days or weeks. | | | |
| A.6.16 | I will have no more than ___ alcohol drinks a night or week. | | | |
| A.6.17 | I will stop consuming caffeinated drinks. | | | |
| A.6.18 | I will stop smoking. | | | |
| A.6.19 | I will stop drinking. | | | |
| A.6.20 | When I have an ailment that I could take an over the counter or prescription drug, I will look to treat what is causing the ailment before taking drugs to ease the symptoms. For example if I have a head ache I will drink more water before taking a Tylenol. | | | |
| A.6.21 | I will re-evaluate when I use recreational drugs and only use them when they serve me the highest, specifically _____. | | | |
| A.6.22 | I will no longer do _____recreational drug because it no longer serves me and who I want to be. | | | |
| A.6.23 | I commit to doing action #(s) ___ consistently for ____ days/weeks/months. | | | |

| | | | | |
|---|---|---|---|---|
| Actions | **Educating and Influencing Others** | | | |
| A.6.24 | I will supply my family or household with healthy food, drink and meals that don't feed our potential addictions or health imbalances. | | | |
| A.6.25 | I will cook my family or friends a meal that limits addictive ingredients. | | | |
| A.6.26 | I will have a discussion with a friend about what I am cutting back on in my life and ask if they would like to join or support me. | | | |
| A.6.27 | I will ask a friend, coworker or family to join me in completing action #(s) _____ from this intention. | | | |

# INTENTION 6 **CUT BACK ON EASY ADDICTIONS**

## EXPLORATIONS 🌍

Sugar and other sweeteners: *http://www.medicinenet.com/artificial_sweeteners/article.htm*

Aspartame: *http://www.ncbi.nlm.nih.gov/pmc/articles/PMC1474447/pdf/envhper00434-0053.pdf*

High fructose corn syrup: *http://www.princeton.edu/main/news/archive/S26/91/22K07/*

Food: The Ultimate Secret Exposed - PT 1/2 *http://www.youtube.com/watch?v=MSpkLk0vYmk*

Earth Save: Promotes healthy and life-sustaining food choices: *www.earthsave.org/about1.html*

Understanding our cravings:

*http://www.bloodtestguide.com/monster-cravings-caused-by-simple-nutrient-deficiencies.html*

Alcohol: *http://www.greenfacts.org/en/alcohol/index.htm*

Tobacco: *http://www.greenfacts.org/en/tobacco/2-tobacco-smoking/4-effects-smoking.htm*

## INTENTION 7 EAT OCEAN WISE

Our ocean and water systems have been an abundant source of food for thousands and thousands of years. These extensive and dynamic ecosystems have undergone many changes in recent generations, and have been seriously affected by our destructive and wasteful fishing practices, by over-fishing the existing stock and by an exponential increase in water pollution. Our decisions now will affect the supply and quality of food for ourselves and future generations and the health of the living ecosystems in our oceans and waterways.

| Actions | The Basics | PLEDGE DATE | # TIMES/ WEEK | # TIMES COMPLETE / 7 WEEKS |
|---------|-----------|---|---|---|
| A.7.1 | When buying fish, I will ask for and buy only wild sustainable fish, rather than farmed fish. 🌍 | | | |
| A.7.2 | Before ordering fish in a restaurant, I will ask if they serve wild sustainable fish and will not order farmed fish. I will choose sustainable wild fish species over unsustainable ones. 🌍 | | | |
| A.7.3 | I will research into sustainable fish at http://www.seachoice.org/ and print a Canada Seafood Guide for my fridge. | | | |

| Actions | I Can Do More | | | |
|---------|--------------|---|---|---|
| A.7.4 | I will carry a SeaChoice Seafood Guide with me when I go to the grocery store or dine out. | | | |
| A.7.5 | I will only eat Seafood rated in the "green" category of the SeaChoice Seafood Guide for ___ days/weeks. | | | |
| A.7.6 | I will get my mercury levels tested, and re-adjust the amount and type of seafood I am consuming depending on the results. 🌍 | | | |
| A.7.7 | I will research into the environmental complexities of shrimp farming in Asia and Latin America before purchasing non-local shrimp. 🌍 | | | |

# INTENTION 7  EAT OCEAN WISE

| | PLEDGE DATE | # TIMES/ WEEK | # TIMES COMPLETE / 7 WEEKS |
|---|---|---|---|

| | | | | |
|---|---|---|---|---|
| A.7.8 | I will research into the environmental complexities of shellfish aquaculture before my next purchase of shellfish. 🌐 | | | |
| A.7.9 | I will learn about the unsustainable fishing practices that are damaging the ocean floor and the high-tech fishing fleets that will have totally consumed all fish stocks within four short decades. Based on this I will cut back on the amount of seafood I eat by ___ %. Enter % that feels best for you. | | | |
| A.7.10 | I will read and sign Alexandra Morton's Petition to protect wild salmon at: www.thepetitionsite.com/305/petition-to-protect-wild-salmon-written-by-alex-andra-morton. | | | |
| A.7.11 | If I eat canned fish I will research into the background and practices used by the companies I buy from. 🌐 | | | |

| | ACTIONS | **Educating and Influencing Others** | | | |
|---|---|---|---|---|---|

| | | | | | |
|---|---|---|---|---|---|
| A.7.12 | I will request my local grocery store to carry and distribute SeaChoice's Canada Seafood Guide and to carry sustainable sea food. | | | |
| A.7.13 | I will request my school or workplace to carry sustainable seafood. I will do research on what their options are and present them to a decision maker. | | | |
| A.7.14 | I will have a happy and abundant sea food pot-luck or dinner to encourage creative ways of enjoying more sustainable dishes. | | | |
| A.7.15 | I will ask a friend, coworker or family to join me in completing action #(s) _____ from this intention. | | | |

| | ACTIONS | **Community Projects** | | | |
|---|---|---|---|---|---|

| | | | | | |
|---|---|---|---|---|---|
| A.7.16 | We will write _____ letters to my local and federal governments regarding our concerns about locally farmed fish or current aquaculture practices, and will provide information and support on more sustainable answers. | | | |
| A.7.17 | We will set up a sustainable seafood information booth to bring to different markets and events. | | | |
| A.7.18 | We will look into the possibility of establishing a community pond specifically for growing and harvesting edible fish. (Also look into aquaponics if you are interested in growing fish and plants together. ) 🌐 | | | |

# EXPLORATIONS

Sustainable fish guides can be found at: *http://www.seachoice.org/*

Mercury levels:

*http://www.hc-sc.gc.ca/hl-vs/iyh-vsv/environ/merc-eng.php http://www.nrdc.org/health/effects/mercury/guide.asp*

Ranking supermarkets in their stand on sustainable seafood:

*www.greenpeace.org/canada/en/recent/Canadian-Supermarkets-Driving-Change-on-Seafood-Sustainability-Greenpeace/*

Shrimp Farming:

*http://news.nationalgeographic.com/news/2004/06/0621_040621_shrimpfarm.html*

*http://www.mongabay.com/environmental_degradation_shrimp.htm*

## INTENTION 7 **EAT OCEAN WISE**

Shellfish aquaculture:
*http://www.responsibleshellfishfarming.ca/environ.htm*
*http://www.pcsga.org/pub/uploads/EnvironShellfishBibliography.pdf*
Canned fish:
*www.greenpeace.org/canada/en/campaigns/Seafood/Get-involved/whats-in-tuna-can/*
Growing own fish:
*http://www.motherearthnews.com/Sustainable-Farming/1970-03-01/Grow-Your-Own-Fish.aspx*
*http://www.theglobeandmail.com/life/food-and-wine/my-backyard-fish-farm/article1218280/*

## INTENTION 8 EAT HAPPY MEAT

It is important to recognize the process and treatment that is behind the meat that we eat. What are you condoning and supporting when you buy that juicy cheap sirloin steak? Various books and documentaries have exposed the conditions in factory farms now used to raise animals for meat, dairy or eggs. The majority of the meat we eat has been factory farmed; 99% of American meat products come from a factory farm.[9] Once exposed, the ethics behind large-scale meat production is scary and alarming. It is time to ask some tough questions about the conditions under which our meat is raised and harvested, and take some ownership in our consumer choices. The actions below can help align the value we place on our health with our commitment to an Earth and animal friendly lifestyle.

| Actions | The Basics | PLEDGE DATE | # TIMES/ WEEK | # TIMES COMPLETE / 7 WEEKS |
|---|---|---|---|---|
| A.8.1 | I will purchase free range eggs. | | | |
| A.8.2 | I will buy free range poultry. | | | |
| A.8.3 | I will buy grass-fed beef. | | | |
| A.8.4 | I will learn about animal conditions in modern agricultural industry. Example: www.peta.org | | | |
| A.8.5 | I will watch the video The Meatrix, (www.TheMeatrix.com) to learn more about factory practices. | | | |

| Actions | I Can Do More | | | |
|---|---|---|---|---|
| A.8.6 | I will support small, community farms with SPCA (ASPCA) approval rather than purchasing meat produced in factory farms. | | | |
| A.8.7 | I will ask questions about how the meat I purchase is produced Example: If antibiotics are used, what farming practices are used, how the animals are raised. A printable Glossary of Meat Production is available at: www.sustainabletable.org | | | |
| A.8.8 | I will support the ethical treatment of animals by joining PETA. | | | |

9. www.farmforward.com/farming-forward/factory-farming

# INTENTION 8  **EAT HAPPY MEAT**

| | | PLEDGE DATE | # TIMES / WEEK | # TIMES COMPLETE / 7 WEEKS |
|---|---|---|---|---|
| A.8.9 | I will check out my local slaughter house(s) and research where their meat or poultry comes from, then goes to; tracing my meat back to its sources from production, noting their treatment and facilities.  | | | |
| A.8.10 | I commit to doing action #(s) ____ consistently for ____ days/weeks/months. | | | |

| | | | | |
|---|---|---|---|---|
| Actions | **Educating and Influencing Others** | | | |
| A.8.11 | I will talk to ___ people about the problems with industrial agriculture and the advantages of sustainable food. Presentation kits are available at www.sustainabletable.org | | | |
| A.8.12 | I will invite friends over for a "happy meat" dinner. | | | |
| A.8.13 | I will ask a friend, coworker or family to join me in completing action #(s)_____ from this intention. | | | |

## EXPLORATIONS 🌐

SPCA Certified Humane Certificate Program:

An inspection, certification, and labeling program designed to ensure that animals raised for food are treated humanely throughout the entire production process.

*http://www.aspca.org/fight-animal-cruelty/farm-animal-cruelty/what-is-certified-humane.html*

*The Omnivore's Dilemma,* by Michael Pollan, Penguin Press, 2006.

*Earthlings Movie: www.earthlings.com*

Meat production, land use and natural resources – the facts behind our diets:

*www.peta.org/issues/animals-used-for-food/meat-wastes-natural-resources.aspx*

*The Way We Eat: Why our Food Choices Matter,* by Peter Singer and Jim Mason, Rodale Books, 2006.

Become informed about the food you eat, and how you can eat more sustainably, by referring to online websites and magazines, such as *http://www.sustainabletable.org/home.php* or *"The Meatrix" www.TheMeatrix.com*

Factory Farming:

*www.farmforward.com/farming-forward/factory-farming*

*www.sustainabletable.org/issues/factoryfarming/*

Free Range Meat:

*www.organicmeat.co.za/freerangebenefits.html*

*www.organicmeat.co.za/freerange.html*

*"It is no measure of health to be well adjusted to a profoundly sick society."*

– J. Krishnamurti, author, speaker, and philosopher

## INTENTION 9 EAT LESS MEAT

There are many reasons that people reduce the amount of meat they are eating or cut it out all together. Some are just drawn to this diet instinctively. Others make this choice as a conscious action to support their priorities and values. Meat-based diets require huge inputs of energy and water. It takes more than 2,400 gallons of water to produce 1 pound of meat, while growing 1 pound of wheat only requires 25 gallons. You save more water by not eating a pound of meat than you do by not showering for six months! A totally vegan diet requires only 300 gallons of water per day, while a typical meat-eating diet requires more than 4,000 gallons of water per day. It takes more than 11 times as much fossil fuel to make one calorie from animal protein as it does to make one calorie from plant protein.[10] The average North American meat-based diet generates approximately 1.5 tonnes of greenhouse gas emissions more per year than vegetarian diets. A mostly (or entirely) vegetarian diet based on whole foods can be gentler both on our bodies and the planet, especially if the food is grown or produced organically, using sustainable agricultural practices. There are plenty of good, plant-based sources of protein (tofu, nuts, beans and lentils) that make delicious, filling meals and for less money than meat-based dishes. Have fun expanding your repertoire of tasty recipes.

| Actions | The Basics | PLEDGE DATE | # TIMES/ WEEK | # TIMES COMPLETE / 7 WEEKS |
|---|---|---|---|---|
| A.9.1 | I will get a vegetarian cookbook and try out ___ new meat free recipe(s) this week. | | | |
| A.9.2 | I will learn more about eating healthy with less or no meat through www.earthsave.ca or another organization in my area. | | | |
| A.9.3 | I will substitute poultry for beef. 🖉 | | | |
| A.9.4 | I will have a vegetarian meal instead of eating poultry. 🌐 🖉 | | | |
| A.9.5 | I will have a vegetarian meal instead of eating meat. 🌐 🖉 | | | |
| A.9.6 | I will order a vegetarian meal next time I dine at a restaurant. | | | |

| Actions | I Can Do More | | | |
|---|---|---|---|---|
| A.9.7 | I will devote ___ hours to menu planning, and buy vegetarian foods that are healthy both to eat and produced via sound agricultural practices. Example: Free range eggs, organic local vegetables and fruit. | | | |
| A.9.8 | I will learn about hormones and antibiotics fed to animals that are then butchered for human consumption. | | | |
| A.9.9 | Based on my values and ethical concerns, I will not eat meat ___ days, and will consider making this a regular weekly practice throughout the year. | | | |
| A.9.10 | I will learn about proteins within a vegetarian diet. | | | |

10. http://www.peta.org/issues/animals-used-for-food/meat-wastes-natural-resources.aspx

# INTENTION 9  **EAT LESS MEAT**

| | | PLEDGE DATE | # TIMES / WEEK | # TIMES COMPLETE / 7 WEEKS |
|---|---|---|---|---|
| A.9.11 | I will learn about vegan nutrition and try a vegan diet (no animal or animal bi-products) for __ days. | | | |
| A.9.12 | I commit to doing action #(s) ____ consistently for ____ days/weeks/months. | | | |

| ACTIONS | **Educating and Influencing Others** | | | |
|---|---|---|---|---|
| A.9.13 | At my school or work cafeteria, I will suggest they offer a vegetarian option for each meal. | | | |
| A.9.14 | At my school or work cafeteria, I will suggest that we have one day a week that is meatless like "Meatless Mondays." | | | |
| A.9.15 | I will make ___ balanced vegetarian meals for my family each week. | | | |
| A.9.16 | I will invite my family, friends, co-workers or neighbors over for a tasty balanced vegetarian meal. | | | |
| A.9.17 | I will ask a friend, coworker or family to join me in completing action #(s)_____ from this intention. | | | |

## EXPLORATIONS

Meat production, land use and natural resources – the facts behind our diets:
www.peta.org/issues/animals-used-for-food/meat-wastes-natural-resources.aspx
Food Security: www.foodsecurecanada.org
World Food Security: Food and Agriculture Organizations (FAO) of the United Nations: www.fao.org
Learn about vegetarianism at Go Veg: http://www.goveg.com/
Earth Save: Promotes healthy and life-sustaining food choices: www.earthsave.org/about1.html

# INTENTION 10  THE ART OF COMPOSTING

Composting is a great way to cut down on solid wastes that end up in landfills and contribute greatly to the emission of greenhouse gases such as methane.[11] With some exceptions, most food scraps are compostable. The end-product of composting - humus, is great for gardens. Lawn clippings and other garden debris can also be used as mulch or ground cover in the garden. Composting is good for the planet, relatively easy on your wallet and great for you or your neighbours plants. So enjoy taking on this win-win action, and reducing your impact everyday!

| ACTIONS | **The Basics** | | | |
|---|---|---|---|---|
| A.10.1 | I will visit an urban agriculture demonstration centre this week to learn more about backyard composting. In the Metro Vancouver area, see SPEC. | | | |

11. Landfills are responsible for 38% of Canada's methane emissions. Greenpeace Green Living Guide (2007), p.142

# INTENTION 10  THE ART OF COMPOSTING

| | | PLEDGE DATE | # TIMES/ WEEK | # TIMES COMPLETE / 7 WEEKS |
|---|---|---|---|---|
| A.10.2 | I will install a backyard or indoor composter this week. Contact your local municipality or region to see what they may offer. 🌐 | | | |
| A.10.3 | I will keep all my food scraps in a plastic bag in the freezer, and put them into my composter or into the leaf & lawn clippings picked up by my municipality. | | | |
| A.10.4 | I will leave the grass clippings on the lawn this week. Grass clippings act as a natural fertilizer, reduce evaporation of water, and reduce the need to water my lawn. 🌐 | | | |
| Actions | **I Can Do More** | | | |
| A.10.5 | I will sharpen my mower blades, and keep them in a "high" position when mowing my lawn. Mowing the lawn at a higher height of 5 to 6cm (2 inches) provides shade to protect your lawn from heat. Deeper roots develop to crowd out weeds and allow your lawn to store more water. | | | |
| A.10.6 | I will provide mulch (nature's composting system of leaves and decomposing materials) or ground cover for my garden. | | | |
| A.10.7 | I will learn about keeping worms as a composter system. Especially helpful if living in an apartment. Contact your local municipality or region for information. 🌐 | | | |
| Actions | **Educating and Influencing Others** | | | |
| A.10.8 | I will ask a friend, coworker or family to join me in completing action #(s)_____ from this intention. | | | |
| Actions | **Community Projects,** | | | |
| A.10.9 | If we live in apartment buildings, we will request that our recycling services include organic matter pickup. | | | |
| A.10.10 | We will start a community composting program. | | | |
| A.10.11 | We will start a composting program for organic waste in the kitchen(s) of our schools and/or workplaces. | | | |

## EXPLORATIONS

Composting information: *www.ec.gc.ca/education/default.asp?lang=En&n=BAE2878A-1*

or *www.metrovancouver.org/services/solidwaste/Pages/default.aspxaspx*

Contact your local municipality to learn of composting programs and to purchase a composter or worm bin.

In BC call the *Compost Hotline at 604-736-2250*

City Farmer /workshops: *http://www.cityfarmer.org/*

Lawn Tips: *www.metrovancouver.org/about/publications/Publications/LawnSprinking.pdf*

Lawn clippings and water wise use: *www.metrovancouver.org/about/publications/Publications/WaterwiseGardening.pdf*

Natural yard care: *www.metrovancouver.org/about/publications/Publications/NaturalYardCareBooklet.pdf*

## *INTENTION 11 TOXIN FREE GARDENING

We like to see our gardens—whether full-size back yards or potted plants on a balcony—looking healthy and flourishing. As part of the gardening process, weeds and pests need to be dealt with to prevent them from taking over our gardens. It may be tempting to opt for convenience and simply apply synthetically-produced herbicides and pesticides, but these products are often highly toxic and take their toll on the environment, on songbirds, butterflies, wildlife, on humans and pets.[12] Similarly, it is important that we find non-toxic approaches to removing pests from inside our homes. If you are able to eat vegetables, shortly after they are picked ensures that you receive all the vitamins and nutrients from them. Many purchased fruits and vegetables are picked prematurely, and treated chemically to ripen and give them a longer shelf life. This process not only makes the produce less tasty but also affects the nutritional content. [13]

| Actions | The Basics | PLEDGE DATE | # TIMES/ WEEK | # TIMES COMPLETE / 7 WEEKS |
|---|---|---|---|---|
| A.11.1 | I will grow organic fruit and vegetables for my own consumption. If I don't have space for a garden plot, I will try growing food that requires less space, like upside-down tomatoes, or use planters, a condo balcony or roof, window cells, and indoor pots. | | | |
| A.11.2 | I will buy only biological pesticides or use plants that repel specific insects. | | | |
| A.11.3 | I will install a lady-bug house and ladybugs for my garden. | | | |
| A.11.4 | I will avoid toxic chemicals when dealing with indoor pests. | | | |
| A.11.5 | I will buy organic potting soil, seeds, fertilizer, and pest and weed control products for future garden projects. | | | |
| A.11.6 | I will buy a garden share in the produce grown by an urban or rural organic farm. | | | |

| Actions | I Can Do More | | | |
|---|---|---|---|---|
| A.11.7 | This week, I will replace lawn and garden chemical pesticides with simple, environmentally-friendly pest controls, such as lining flowerbeds and plants with eggshells, using chickens to remove pests, or dog hair to protect plants from slugs. | | | |
| A.11.8 | I will turn my yard into a balanced ecosystem. I will aim to create a low maintenance yard, with a wide variety of local plant species that require less water. I will plant ___ local plant species. | | | |
| A.11.9 | I will visit ___ community gardens for inspiration and advice this week. | | | |
| A.11.10 | I will start my own greenhouse for cultivating vegetables. Try to purchase one second-hand through Craigslist. | | | |
| A.11.11 | I will check out local colleges, community and garden centres for a class on gardening and register for one. 🌐 | | | |
| A.11.12 | I will start cultivating medicinal herbs and other herbs for cooking. | | | |

12. According to the Sierra Club of Canada, more than 34 million kilograms of pesticides are used in Canada per year. www.sierraclub.ca/national/programs/health-environment/pesticides/index.shtml
13. http://www.livestrong.com/article/333160-the-advantages-of-fresh-vegetables-picked-from-the-garden/#ixzz1Sa3zZUJZ

# INTENTION 11 TOXIN FREE GARDENING

| | | PLEDGE DATE | # TIMES/ WEEK | # TIMES COMPLETE / 7 WEEKS |
|---|---|---|---|---|
| A.11.13 | I will keep informed by using the internet or organic encyclopedias to cultivate plants. | | | |
| A.11.14 | I will switch to manual, electric or even solar powered lawn equipment if I currently use gas-powered equipment in my garden. | | | |
| A.11.15 | I commit to doing action #(s) ____ consistently for ____ days/weeks/months. | | | |
| ACTIONS | **Educating and Influencing Others** | | | |
| A.11.16 | I will share seeds with friends and neighbours. | | | |
| A.11.17 | I will invite others over to share my harvest. | | | |
| A.11.18 | I will invite family, friends, coworkers or neighbours to come over and garden with me. | | | |
| A.11.19 | I will ask a friend, coworker or family to join me in completing action #(s)_____ from this intention. | | | |
| ACTIONS | **Community Projects,** | | | |
| A.11.20 | We will advocate that our schools and/or workplaces be pesticide and herbicide-free, providing the decision makers with usable actions and research which to base their decisions on. | | | |
| A.11.21 | We will get involved in or create a neighbourhood gardening circle or join a community garden to grow food and herbs.  | | | |
| A.11.22 | We will apply to convert vacant city plots into community gardens. | | | |
| A.11.23 | We will start community gardening/produce shares within our neighbourhoods. | | | |

# EXPLORATIONS

Rule of-thumb: don't overreact, 1 or 2 pests inside are not an invasion; locate and block pests' point of entry; keep kitchen, floors, and garbage pails clean to eliminate pests' food supplies; and remove clutter to eliminate nesting areas.

City gardening around the world: *www.cityfarmer.info*

Organic gardening: *www.organicgardeningresources.com/or www.organicgardening.com/*

Most cities have a multitude of gardening workshops and demonstrations. Check online for the ones nearest you.

Natural gardening tips: *http://www.evergreen.ca/en/resources/home-greening/gardening-tips.sn*

Gardening workshops and demonstrations: *www.spec.bc.ca/gardens*

Cultivating Food and Community in BC, Village Vancouver, A Fork In The Road: *www.villagevancouver.ca*

Two Block Diet: *http://bit.ly/qLy9Dw*

Get your yard off drugs: *www.vancouver.ca/engsvcs/solidwaste/grownatural/offdrugs.htm*

Learn about the process of proposing and creating a community garden in the City of Vancouver at Vancouver's Food Policy Community Garden Resources: *www.vancouver.ca/commsvcs/socialplanning/initiatives/foodpolicy/projects/gardenresource.htm*

Chemical free gardens: *http://www.davidsuzuki.org/publications/downloads/2008/DSF_Pesticide_Free_Oui_En.pdf*

| | PLEDGE DATE | # TIMES / WEEK | # TIMES COMPLETE / 7 WEEKS |

## INTENTION 11  TOXIN FREE GARDENING

For information on safe pest removal: *http://watoxics.org/healthy-living/healthy-homes-gardens-1*
*www.sierraclubgreenhome.com/go-green/landscaping-and-outdoors/safe-pesticides/*
*www.sierraclubgreenhome.com/go-green/pest-control-indoors/safe-indoor-pest-control/*
For pesticide information on alternatives and disposal: *http://www.crd.bc.ca/gardening/*
How and why pesticides are harmful, Putting your yard and garden into de-tox:
*www.georgiastrait.org/files/share/PDF/TSpesticides.pdf*
Safe disposal of your pesticides and herbicides in BC: *http://rcbc.bc.ca/*
For western Canada see *www.productcare.org* or contact your local recycling hotline or recycling centre or municipality.

## INTENTION 12  SAY NO TO TOXINS

A report by the Columbia University School of Public Health estimated that 95 % of cancer is caused by our diets and environmental toxicity. Most North Americans have somewhere between 400 and 800 chemicals stored in their bodies, typically stored in their fat cells. Toxins are everywhere, in our water, air and in our food. Most often we don't realize that we are being affected. But after extended, consistent and subtle exposure to a multitude of toxins in our environments, products and food the result is often a chronic disease.[14] Organized laborers are starting to control their exposure to harsh chemical cleaners or other toxins that pollute indoor air quality; we too can control the exposure we personally endure. From cosmetics, to toiletries, medicine, clothing and electronics, we can choose to eliminate the use of chemically-based (usually from petroleum by-products) products and their hazardous consequences from our personal use, and ultimately from the environment. We can opt for certified eco-friendly products, and, better yet to save money, reduce our overall consumption. When we live as toxin and chemical free as possible, we are kinder to our health and the planet.

| Actions | The Basics | | | |
|---|---|---|---|---|
| A.12.1 | I will purchase toxin-free, natural personal-care products and toiletries. | | | |
| A.12.2 | I will research toxins in my cosmetic make-up at Cosmetic Safety Database www. cosmeticsdatabase.com, and make healthier choices using toxic free cosmetics. | | | |
| A.12.3 | I will go to a natural dry cleaner that does not use toxic solvents. | | | |
| A.12.4 | I will explore alternatives to conventional pharmaceuticals to meet my health-care needs, such as herbal, homeopathic, natural and bio-identical products and supplements. | | | |
| A.12.5 | I will not throw any pharmaceuticals or medicines down the toilet. I will recycle at my local pharmacist store all my expired or un-used pharmaceuticals or medicines including: all prescription drugs, all non-prescription medicines, mineral supplements, vitamin supplements, and throat lozenges. This will reduce further toxins from entering our environment. 🌍 | | | |
| A.12.6 | I will avoid clothing with fabrics derived from petrochemicals, or treated with chemicals and dyes that pollute our environment. 🌍 | | | |

14. www.scmhealth.com/content/effects_toxins.html

# INTENTION 12 **SAY NO TO TOXINS**

| | | PLEDGE DATE | # TIMES/ WEEK | # TIMES COMPLETE / 7 WEEKS |
|---|---|---|---|---|
| ACTIONS | **I Can Do More** | | | |
| A.12.7 | I will avoid purchasing items that require persistent chemicals. | | | |
| A.12.8 | I will buy my next electronics purchase from manufacturers that are committed to helping the environment. | | | |
| A.12.9 | I will explore the issue of vaccine toxicity prior to receiving vaccinations. 🌐 | | | |
| A.12.10 | I will limit my dry cleaning by ___items this week and hand wash items I would otherwise have dry-cleaned. | | | |
| A.12.11 | I will avoid purchasing clothing that requires dry-cleaning. | | | |
| A.12.12 | I will investigate the use of organo-chlorines in the bleaching process of feminine hygiene products at www.emagazine.com, and will choose to make sustainable health choices. 🌐 | | | |
| A.12.13 | I commit to doing action #(s) ____ consistently for ____ days/weeks/months. | | | |
| ACTIONS | **Educating and Influencing Others** | | | |
| A.12.14 | I will share the knowledge with my friends or family when I find a good toxic or chemical free product, company or service. | | | |
| A.12.15 | When I buy a gift for a friend or family member, I will let them know that I chose a toxic-free product to keep them and the environment healthy. | | | |
| A.12.16 | I will ask a friend, coworker or family to join me in completing action #(s)_____ from this intention. | | | |
| ACTIONS | **Community Projects,** | | | |
| A.12.17 | We will support the government in banning known toxic substances. | | | |
| A.12.18 | We will advocate for the "Precautionary Principle" in the regulation of chemicals (if there is a weight of evidence that a chemical is carcinogenic or toxic, there should be regulations restricting its use or even banning it). | | | |
| A.12.19 | We will insist that governments introduce "Right-to-Know" labeling requirements on manufacturers to disclose what is in their products. | | | |
| A.12.20 | We will personally implement action #(s)_____ from this intention at our schools and/or workplaces. | | | |
| A.12.21 | We will request that our schools and/or workplaces consider implementing action #(s) _____ from this intention; we will research how this is done to make the process as easy as possible. | | | |

# INTENTION 12 **SAY NO TO TOXINS**

| | PLEDGE DATE | # TIMES/ WEEK | # TIMES COMPLETE / 7 WEEKS |
|---|---|---|---|

## EXPLORATIONS 🌍

Read: *The Greenpeace Green Living Guide (2007), pages 40-44.*

Government of Canada:

Fire Retardants (Polybrominated Diphenyl Ethers PBDEs) which are used to slow the spread of fire in a variety of plastics and other products. PBDEs are toxic to our environment because they build up over time and last a long time in our environment.

*http://www.ec.gc.ca/default.asp?lang=En&n=714D9AAE-1&news=C0D6EB45-514A-4050-98B5-9B8A6730BBFB*

Recycling your prescription and non-prescription pharmaceutical medicines in Canada:
*http://www.medicationsreturn.ca/*

Research green alternatives the next time you go to buy a computer or other electronic device:
*http://www.greenpeace.org/usa/en/news-and-blogs/news/green-gadgets-the-search-con/*

Health Canada's safe use of cook ware: *www.hc-sc.gc.ca/hl-vs/iyh-vsv/prod/cook-cuisinier-eng.php*

Federal Government of Canada, Health Canada, Aluminium and your health:
*http://www.hc-sc.gc.ca/fn-an/securit/addit/aluminum-eng.php*

Toxic Free Canada is working with other groups and campaigning for a Canadian national "Right-to-Know" legislation that would require ingredient and hazard labelling for household products:
*www.toxicfreecanada.ca/campaign.asp?c=2*

Vaccine Toxicity:
*http://www.healthychild.com/vaccine-choices/vaccine-toxicity-and-safety-of-vaccinations-a-parents-right-to-choose/*

Inform yourself about the ways in which we can remove toxins from our products. Learn about the Extended Producer Responsibility Law (EPR) at the following sites:
*http://www.unep.fr/en/ and http://en.wikipedia.org/wiki/Extended_producer_responsibility*

The Greenpeace Green Living Guide has well researched and concise info on: electronics: p. 54–58

# INTENTION 13  A TOXIC FREE HOUSEHOLD

Gardening and household cleaning products are not the only environmentally damaging toxic substances in and around our homes. Items as diverse as electronics, flea collars for pets, oil-based paints and finishers, adhesives, synthetic fabrics, and floor covering material (e.g., materials used for carpets, synthetic linoleum), may all include toxic substances - either as specific ingredients or as part of the manufacturing process, or be produced from non-renewable resources through environmentally damaging methods. Choose to follow up on the actions below to gradually eliminate as many toxins as possible from your home and environment.

| Actions | The Basics | | | |
|---|---|---|---|---|
| A.13.1 | I will purchase eco-friendly cleaning products, be aware of misleading labels, and look for a credible third party certification label. | | | |
| A.13.2 | I will use vinegar to clean glass, and other natural products. I was able to replace ___ toxic cleaning products with natural alternatives this week. | | | |
| A.13.3 | I will stop using household products labeled poisonous, explosive, corrosive or flammable, and all other toxic cleaning products that I have been using. | | | |

# INTENTION 13  A TOXIC FREE HOUSEHOLD

| | | PLEDGE DATE | # TIMES/ WEEK | # TIMES COMPLETE / 7 WEEKS |
|---|---|---|---|---|
| A.13.4 | I will check out www.greenhome.com/products. | | | |
| A.13.5 | I will avoid and refrain from using Teflon or non-stick and aluminum cooking ware. 🌍 | | | |
| A.13.6 | I will only purchase children's clothing made with natural fire-retardants. | | | |
| A.13.7 | If my pet wears a flea collar, I will find the safest product. | | | |
| A.13.8 | I will limit my use of paints, finishers, and adhesives. | | | |
| A.13.9 | When it is necessary to paint, I will dispose of paint products carefully. I will contact my local municipality about recycling programs and disposal instructions. | | | |
| A.13.10 | When buying paints, I will try to use only latex paint. I will look for Green Seal products that have low Volatile Organic Compounds (VOC) levels, and purchase only as much as I will need. | | | |

| ACTIONS | I Can Do More | | | |
|---|---|---|---|---|
| A.13.11 | I will explore safer alternatives for in-home furnishings, and choose furnishings (such as mattresses and pillows) that have natural fire-retardants. 🌍 | | | |
| A.13.12 | I will avoid using toxic home construction materials during new-home construction or renovations or repairs. Example: research alternatives for PVC (vinyl), toxic paints, etc. | | | |
| A.13.13 | I will research ___ number of purchased good(s) to see if its contents are non-toxic. For example computer chairs, toilet paper, mattresses, food stuffs, plastic bottles, etc. | | | |
| A.13.14 | I will choose floor coverings made from renewable resources or natural materials. | | | |
| A.13.15 | I will find and use carpet cleaning businesses that use natural cleaners. | | | |
| A.13.16 | I commit to doing action #(s) ___ consistently for ___ days/weeks/months. | | | |

| ACTIONS | Educating and Influencing Others | | | |
|---|---|---|---|---|
| A.13.17 | I will buy and research the Cancer Smart Consumer Guide: http://www.leas.ca/CancerSmart-Consumer-Guide.htm to develop a list of preferred environmentally friendly products that we can use (cleaners, toilet paper, soaps, etc.). I will bring my research of preferred environmentally friendly products to a decision-maker at work. | | | |
| A.13.18 | I will share my finding from action #____ with a friend or family member. | | | |
| A.13.19 | I will ask a friend, coworker or family to join me in completing action #(s)____ from this intention. | | | |

# INTENTION 13  **A TOXIC FREE HOUSEHOLD**

| | | PLEDGE DATE | # TIMES/ WEEK | # TIMES COMPLETE / 7 WEEKS |
|---|---|---|---|---|
| Actions | **Community Projects,** | | | |
| A.13.20 | We will lobby businesses and corporations to use safer materials in the manufacture of their products. We will enquire about what is in the product, and refuse to buy it if there is a health risk to ourselves, the people who made it, or the environment. | | | |
| A.13.21 | We will personally implement action #(s)_____ from this intention at our schools and/or workplaces. | | | |
| A.13.22 | We will request that our schools and/or workplaces consider implementing action #(s) _____ from this intention; we will research how this is done to make the process as easy as possible. | | | |

## EXPLORATIONS

Read: *The Greenpeace Green Living Guide (2007), pages 40-44.*

Learn about efforts to get toxics out of homes and workplaces: *www.davidsuzuki.org/health*

For more information on which cleaners are toxic: *www.globalstewards.org/toxics.htm*

Research how to clean with safe alternatives and natural disinfectants: *http://www.georgiastrait.org/?q=node/371*

Research the concerns regarding flooring made from polyvinyl chloride at *www.pvcinformation.org*

PVC (Vinyl) Health Concerns: *http://www.greenpeace.org/usa/en/media-center/reports/pvc-the-poison-plastic/*

Learn about toxins in your household products at Toxic Free Canada, *www.toxicfreecanada.ca*

PFOA (perfluorooctanoic acid) and PFOS (perfluorooctane sulfonate) chemical formulations are the original primary constituents of Teflon and Scotchguard products and are potentially toxic.

Read about this toxicity to humans and our environment, see:

*www.ewg.org/node/8732 and www.ewg.org/news/nonstick-chemicals-linked-infertility*

Learn about what you can do about Teflon / non-stick cooking ware / anodized aluminium: *http://www.cancer.ca/ Canada-wide/Prevention/Whats%20being%20studied/Teflon%20and%20non-stick%20cookware/Tips%20to%20reduce%20your%20exposure%20to%20Teflon-related%20substances.aspx?sc_lang=en*

The Greenpeace Green Living Guide has well researched and concise info on:, on floor coverings p. 72–76 and on paints p. 92–94.

*"There is only one thing more powerful than all the armies of the world, that is an idea whose time has come."* – Victor Hugo

# INTENTION 14 REDUCE EFFECTS OF RADIATION

The times are changing. And the things we need to be aware of, for our personal health, are also constantly changing. The earthquake in Japan in March, 2011 has affected radiation levels all over the world. We have to be aware of the personal consequences of using Nuclear energy. What are the steps we can take to protect ourselves and our families? America also has nuclear energy plants that are vulnerable to rising rivers (from climate change) and the unpredictable but recurring incidence of natural disasters. This increasing threat of exposure to harmful radiation is a serious concern in our world. The following steps are recommended to help mitigate serious health effects from radiation. And whenever your voice can be heard, we encourage everyone to influence the stop of all nuclear power plants.

| Actions | The Basics | PLEDGE DATE | # TIMES/ WEEK | # TIMES COMPLETE / 7 WEEKS |
|---|---|---|---|---|
| A.14.1 | I will educate myself on the affects of a nuclear fallout. 🌍 | | | |
| A.14.2 | I will educate myself on how I can eat to mitigate the effects of radiation in my body. 🌍 | | | |
| A.14.3 | I will start eating more _____ and _____ to help my body handle the effects of radiation. 🌍 | | | |
| A.14.4 | I will start to take the appropriate supplements or minerals for me to mitigate the effects of radiation. 🌍 | | | |
| A.14.5 | I will advocate to stop the use of nuclear power and weapons by signing appropriate petitions. 🌍 | | | |
| A.14.6 | I will research into where my water comes from. Radiation stays more present in bodies of water held by a lake or other form of open containment, and less in fast moving water or filtered through the water table. | | | |

| Actions | I Can Do More | | | |
|---|---|---|---|---|
| A.14.7 | I will source the best current and local advisories that monitor air and water contamination that I can find, (while recognizing that the monitors/geiger counters available to citizens don't catch the particles, eg. Plutonium, strontium and americium) and take additional precautions when the levels become higher. 🌍 | | | |
| A.14.8 | I will urgently demand that my government provide widespread and continuous air and water monitoring of radiation levels. | | | |
| A.14.9 | I will be careful to use rain gear when going into the rain, and will leave wet shoes and outerwear in a contained area when levels are higher. | | | |
| A.14.10 | I will take an epsom salt bath ___ times a week/month (consider adding a tsp of baking soda). | | | |
| A.14.11 | I will become aware of the amount of fish and ocean products I eat from the Pacific Ocean, and the amount of big leaf greens I eat from the West Coast of North America, to eat to counteract the affects of radiation. | | | |

# INTENTION 14  REDUCE EFFECTS OF RADIATION

| | | PLEDGE DATE | # TIMES/ WEEK | # TIMES COMPLETE / 7 WEEKS |
|---|---|---|---|---|
| A.14.12 | I will urgently request from my government the testing of food from the sea for nuclear and other contamination. | | | |
| A.14.13 | I will try to eat more root vegetables and less big leafy greens if I live on the West Coast of North America. | | | |
| A.14.14 | I will try to eat more fish from moving rivers versus the Pacific Ocean or lakes. | | | |
| A.14.15 | I will build a greenhouse for my leafy greens and water them with water from the tap not a rain barrel when contamination levels are high. | | | |
| A.14.16 | I commit to doing action #(s) ____ consistently for ____ days/weeks/months. | | | |

| | | | | |
|---|---|---|---|---|
| ACTIONS | **Educating and Influencing Others** | | | |
| A.14.17 | I will start a dialogue about the effects of nuclear radiation once I have done some research myself. | | | |
| A.14.18 | I will email share the results from tracking the radiation levels/fallout in local areas, and recommend additional precautions when the levels become higher. 🌐 | | | |
| A.14.19 | I will inform any of my friends or family that are pregnant or have young children to be particularly aware of what to avoid and send them resources to inform them on what to avoid. | | | |
| A.14.20 | I will write a letter about how we can protect ourselves and simple facts about the effects of nuclear radiation and send it to friends and family, and to a local paper. | | | |
| A.14.21 | I will ask a friend, coworker or family to join me in completing action #(s) _____ from this intention. | | | |

| | | | | |
|---|---|---|---|---|
| ACTIONS | **Community Projects,** | | | |
| A.14.22 | We will join local organizations and/or organize local action circles to garner and evaluate information on monitoring radiation levels in the air, soil, food and water. | | | |
| A.14.23 | We will educate ourselves about the inseparable connection between nuclear power and nuclear weapons production. | | | |
| A.14.24 | We will join local organizations and/or organize local study action circles to advocate stopping the use of nuclear power and weapons. | | | |
| A.14.25 | We will write and circulate a petition to the government to support water sourcing that would be less affected by radiation. | | | |
| A.14.26 | We will put pressure on our governments to do continual and ubiquitous radiation monitoring of air, soil, food and water, and to provide us with health advisories. | | | |
| A.14.27 | We will sign and send out a petition to stop any more nuclear power plants to be built in our country or at ____ location, as we are all interconnected and feel the effects of these decisions. | | | |

# INTENTION 14 **REDUCE EFFECTS OF RADIATION**

## EXPLORATIONS

Nuclear Energy, Nuclear Weapons, and Renewable Energy Sources:

*http://transitionnow.wordpress.com/2011/04/23/radiation-fallout-from-japan-affects-food-safety-across-north-america/* Particularly read Specific Foods and Geographic Areas of Concern section of that page, as it has a nice overview.

*http://www.beyondnuclear.org/*

*http://www.nirs.org/home.htm*

*http://www.ieer.org/*

*http://www.ieer.org/carbonfree/index.html*

*http://www.psr.org/*

*http://fairewinds.org*

*http://trivalleycares.org*


Eating for mitigating the effects of radiation:

*http://www.waccobb.net/forums/showthread.php?78066-Dr.-Patterson-SUPPLEMENTS-SHOWN-TO-HELP-PREVENT-EFFECTS-OF-RADIATION-FALL-OUT*

*http://midnightapothecary.blogspot.com/2011/03/recipes-for-great-turning.html*

*http://www.joannamacy.net/images/stories/fukushima_japan_2011/0707radiationdetox.pdf*


Tracking radiation levels:

*http://www.endtimesreport.com/jet.html*

*http://radiationnetwork.com/index.htm*


Sign Petitions: (there are so many more than this!)

A Simple Statement On Nuclear Power and Climate Change *http://www.nirs.org/petition2/index.php?r=sb*

Call for the suspension of the 23 Fukushima-style reactors operating in the United States:
*http://www.beyondnuclear.org/home/2011/6/11/support-the-beyond-nuclear-petition-to-nrc-to-suspend-operat.html*

For A Post-Fukushima Program for Increased Nuclear Security and Safety:
*http://org2.democracyinaction.org/o/5502/p/dia/action/public/?action_KEY=6195*


*"The most remarkable feature of this historical moment on Earth is not that we are on the way to destroying the world—we've actually been on the way for quite a while. It is that we are beginning to wake up, as from a millennia-long sleep, to a whole new relationship to our world, to ourselves, and each other."* – Joanna Macy

# VALUE B
# REDUCE MATERIAL CONSUMPTION

*I value longevity, quality over quantity, creative sharing
and spacious simplicity.*

The assumption that 'more is better' is leading to our self-destruction, especially as this relates to material goods.[4] The social norm to constantly upgrade and get the latest products or clothes to suit the newest fashion is part of a cycle that is depleting our resources and filling our landfills, oceans and atmosphere with harmful pollutants.

Most American households produce approximately 4.5 lbs (2 kgs) of solid waste a day, which is 1,642 lbs (730 kgs) a year. And each lb/ kg of waste creates twice as much greenhouse gases![5] The good news is we can choose to disengage from a 'disposable culture' that encourages us to keep madly working to earn more money, to keep madly buying new products, to reach an elusive and illusory goal of happiness and fulfillment. We can choose instead to get creative with our resources and focus more on what brings us meaning and long-lasting happiness. It is time for us to reclaim our identity as citizens instead of consumers.

## I Intend To:

B 1     **Organize Possessions**

B 2     **Manage Money Wisely**

B 3     **Avoid the Lure of Advertising**

B 4     **Revise Spending Habits**

B 5     **Alternatives to Buying New Items**

B 6     **Avoid Disposables**

B 7     **Avoid Using Plastics & Packaging**

B 8     **Reuse Everything**

B 9     **Recycle Everything**

B 10    **Supplement**

4. For more information: Changing Consumption from the United Nations Program: www.unep.org/Documents.Multilingual/
   Default.asp?DocumentID=52&ArticleID=52&l=en
5. Gershon, David. *Low Carbon Diet* (2007), p. 7

# INTENTION 1 ORGANIZE POSSESSIONS

Instead of tripping over our belongings—or being crowded out of our home by material "stuff"— reducing unnecessary belongings and keeping things organized frees up usable space and clears the energy flow in our homes. Less clutter means less time needed to clean and maintain your space, and it is easier and more efficient to find the objects that we need. When we surround ourselves with disarray, we secrete more cortisol in our body, a hormone associated with stress. Clutter is not only a symptom of stress, it is also a cause of stress.[1]

| ACTIONS | The Basics | PLEDGE DATE | # TIMES/ WEEK | # TIMES COMPLETE / 7 WEEKS |
|---|---|---|---|---|
| B.1.1 | I will simplify my wardrobe by getting rid of ___ things I don't actually wear. | | | |
| B.1.2 | I will reduce clutter by spending ___ hours this week boxing up items that I do not use or want to keep for decor. | | | |

| ACTIONS | I Can Do More | | | |
|---|---|---|---|---|
| B.1.3 | I will donate ___ number of boxes of clothing and objects to charity. | | | |
| B.1.4 | I will place in a consignment store, or put on a garage sale, ___ possessions that could be used more productively by others. | | | |
| B.1.5 | Inspired by these ideas, I will take the following action perfectly suited for my lifestyle: _____ | | | |
| B.1.6 | I can personally implement action #(s) _____ from this intention at my school or workplace. | | | |

| ACTIONS | Educating and Influencing Others | | | |
|---|---|---|---|---|
| B.1.7 | I will ask a friend, coworker or family to join me in completing action #(s) _____ from this intention. | | | |

| ACTIONS | Community Projects | | | |
|---|---|---|---|---|
| B.1.8 | We will request that our workplaces and/or schools consider implementing action #(s) _____ from this intention; we will offer our help in this process. | | | |

# EXPLORATIONS

Decluttering Your Home, Leo Babauta's Zen Habits: *www.zenhabits.net*
*www.zenhabits.net/2009/09/my-new-ebook-the-simple-guide-to-a-minimalist-life/*
*www.myhouseandgarden.com/declutter.htm*

1. www.realage.com/tips/the-health-benefits-of-decluttering

# INTENTION 2 MANAGE MONEY WISELY

We live in a society where we are surrounded by retail stores,[2] restaurants, fast-food outlets, instant bank tellers, and other conveniences. This makes it easy to fall into a pattern of impulse buying or opting to eat out, rather than making the effort to pack a lunch or snack from home. Yet, if we get into patterns of unconscious consuming, we find that our money slips away from us all too easily and we end up with nothing to show for the money spent. By choosing to manage money wisely and live well below my means, I will avoid running into debt and am more likely to have financial resources to fall back on when unexpected opportunities or expenses show up. The following actions not only help Earth, they also create more abundance in my life.

| ACTIONS | The Basics | | | |
|---|---|---|---|---|
| B.2.1 | I will reduce my unnecessary eating out by bringing coffee and lunch to work ___ days this week. , dining out less, or hosting a potluck rather than eating out. | | | |
| B.2.2 | I will eat dinner at home __ times this week, instead of dining out. | | | |
| B.2.3 | I will resist impulse buying and record the number of times I was successful. | | | |
| B.2.4 | I will avoid shopping as recreation or therapy and record the number of times I had the urge to go shopping and didn't. | | | |

| ACTIONS | I Can Do More | | | |
|---|---|---|---|---|
| B.2.5 | I will examine my monthly expenditures and find a recording system that helps me to live on a budget. | | | |
| B.2.6 | I will investigate living closer to my place of work to reduce travel and time costs. | | | |
| B.2.7 | I will consider living with a roommate or in co-operative housing, or in a smaller more energy efficient home. | | | |

| ACTIONS | Educating and Influencing Others | | | |
|---|---|---|---|---|
| B.2.8 | I will ask a friend, coworker or family to join me in completing action #(s) _____ from this intention. | | | |

*"Live simply so that others may simply live"* – Gandhi

2. Canadians spent $391.3 billion in retail stores in 2006. Greenpeace Green Living Guide (2007), p.116

# INTENTION 3  AVOID THE LURE OF ADVERTISING

Messages from advertisers to buy all kinds of objects and services surround us. We see them on TV, on billboards, at the movies, in magazines, in newspapers, and we hear them on the radio. Advertisements hook us into subtle wants or insecurities that we may, or may not, be aware of... in order to lure us into buying products that we do not actually need. The following actions help protect me from the lure of advertising and consumer culture.

| ACTIONS | The Basics | PLEDGE DATE | # TIMES/ WEEK | # TIMES COMPLETE / 7 WEEKS |
|---|---|---|---|---|
| B.3.1 | I will ask myself "Do I really need to use or purchase this in order to complete my task or to feel satisfied?" | | | |
| B.3.2 | I will mute the advertisements while watching TV. | | | |
| B.3.3 | I will minimize my exposure to consumer messages by recycling the advertisement inserts before reading the newspaper. | | | |
| B.3.4 | I will think twice about sale pitches that encourage me to buy more than I need, (coupons, two-for-one, buy-now pay-later). | | | |
| B.3.5 | I will avoid viewing store websites on the internet. | | | |
| B.3.6 | I will create a gratitude list at the end of the day, to help me remember the good things and people in my life that bring me happiness. | | | |

| ACTIONS | I Can Do More | | | |
|---|---|---|---|---|
| B.3.7 | I will journal and construct a hierarchy of my own basic needs: those things that are absolutely necessary for me to live, to be fulfilled, and to be happy. *See Value B supplement for more information on Abraham Maslow's Hierarchy of Needs.* | | | |
| B.3.8 | In order to reduce my consumption, I will keep a consumption log/journal and record:<br>• each item purchased<br>• the financial cost<br>• what level of happiness on a scale of 1-10 this item gave me<br>* I will revisit my journal in 2 weeks and estimate my sustained happiness level to assess if these items provided me with ongoing or only momentary happiness. | | | |
| B.3.9 | I will reduce the amount of television my family and children are exposed to (to decrease the amount of adverts they are exposed to). | | | |
| B.3.10 | I will teach my family and children to mute the advertisements when watching television, and have them do this ___ days. | | | |
| B.3.11 | Inspired by these ideas, I will take the following action perfectly suited for my lifestyle: _____ | | | |
| B.3.12 | I commit to doing action #(s) ___ consistently for ___ days/weeks/months. | | | |

## INTENTION 3 **AVOID THE LURE OF ADVERTISING**

| | | PLEDGE DATE | # TIMES/ WEEK | # TIMES COMPLETE / 7 WEEKS |
|---|---|---|---|---|
| ACTIONS | **Educating and Influencing Others** | | | |
| B.3.13 | With a friend, I will make a guessing game of what insecurities, promises and target market each advertisement is focusing on, and "is it working on me?" | | | |
| B.3.14 | I will contact a local radio station about having an hour free of advertisements each day. | | | |
| B.3.15 | I will ask a friend, coworker or family to join me in completing action #(s) _____ from this intention. | | | |
| ACTIONS | **Community Projects** | | | |
| B.3.16 | We will advocate for a public 'Ad Free' day. | | | |

## EXPLORATIONS

Affect of branding products on people:
*www.davidsuzuki.org/about_us/Dr_David_Suzuki/Article_Archives/weekly07290501.asp*

Technology's impact on children:
*www.davidsuzuki.org/about_us/Dr_David_Suzuki/Article_Archives/weekly07280601.asp*

Advertising perspectives: *www.adbusters.org/about/adbusters*

The system and our role in it: *http://www.storyofstuff.com/ See also Supplement to Value B, Intention 3*

## INTENTION 4  REVISE SPENDING HABITS

Part of a sustainable life style involves practicing the "3 R's" of Reduce, Reuse, and Recycle. Revising my spending habits to buy less stuff helps support the first powerful "R," Reduce and becoming a conscious consumer takes some practice. Implementing some of the following actions will help to instill the new habit by asking some crucial questions before making purchases.

| | | PLEDGE DATE | # TIMES/ WEEK | # TIMES COMPLETE / 7 WEEKS |
|---|---|---|---|---|
| ACTIONS | **The Basics** | | | |
| B.4.1 | I will shop with a list and consider the necessity of each desired purchase that is not on my list. | | | |
| B.4.2 | I will only buy what is useful for me and not for an item's status. | | | |
| B.4.3 | I will think critically about whether I truly need an item by practicing 'pre-cycling' before I make a purchase. This entails asking myself four questions: Why do I want this? How will I use it? What are my alternatives? And, can I get along without it? I will practice 'pre-cycling' ___times. | | | |
| B.4.4 | I will create a list of activities to do instead of shopping so that the next time I have an urge to shop, I can easily consider a fun and rewarding alternative to share with a friend (exploring a local natural area, starting a garden or other project, trying out a physical activity). | | | |

## INTENTION 4  **REVISE SPENDING HABITS**

| | | PLEDGE DATE | # TIMES/ WEEK | # TIMES COMPLETE / 7 WEEKS |
|---|---|---|---|---|
| ACTIONS | **I Can Do More** | | | |
| B.4.5 | I will avoid a shopping mall. | | | |
| B.4.6 | I will resist purchasing ___ items that I feel I do not need but that I may be addicted to (coffee, chocolate, books, magazines, clothes, and makeup). | | | |
| B.4.7 | I will give gift certificates for personal services, tickets to an event or something homemade instead of new products for ___ gifts in the next 3 months. | | | |
| B.4.8 | I will carry a small "I don't need it – don't want it" notebook for a full day and write down any item that I feel an impulse to buy, and then I'll keep on walking! The process of slowing down enough to write down the object helps decrease impulse buying. I will look in my notebook and report to my Circle how much stuff I ended up NOT buying. | | | |
| B.4.9 | Inspired by these ideas, I will take the following action perfectly suited for my lifestyle: _____ | | | |
| B.4.10 | I commit to doing action #(s) ____ consistently for ____ days/weeks/months. | | | |

| | | | | |
|---|---|---|---|---|
| ACTIONS | **Educating and Influencing Others** | | | |
| B.4.11 | Next time I am out with a friend that is about to make an impulse buy, I will question them on why they need the item, can they do without it, and encourage them to maybe sleep on the purchase decision. | | | |
| B.4.12 | I will ask a friend, coworker or family to join me in completing action #(s) _____ from this intention. | | | |

| | | | | |
|---|---|---|---|---|
| ACTIONS | **Community Projects** | | | |
| B.4.13 | We will organize an event for 'Buy Nothing' day, http://en.wikipedia.org/wiki/Buy_Nothing_Day | | | |

## INTENTION 5 ALTERNATIVES TO BUYING NEW ITEMS

By choosing to shop second hand for clothes or other objects that we need or want, we are practicing the second "R"- Reuse. In our consumer culture, we are strongly encouraged through social conditioning and advertising to buy new and or disposable goods, rather than "used" goods.[3] Yet, in the long run, choosing to repair, swap, rent or resell existing goods is more sustainable both from an environmental perspective and a wise money management perspective. The following actions are some of the many ways I can avoid making a new purchase when I am need an object.

3. Mass consumption of goods was first aggressively pushed by politicians and business leaders in post-WWII America. Ghazi, Polly and Jones, Judy. Downshifting: The Bestselling Guide to Happier, Simpler Living (2004), p.8

# INTENTION 5 **ALTERNATIVES TO BUYING NEW ITEMS**

| | PLEDGE DATE | # TIMES / WEEK | # TIMES COMPLETE / 7 WEEKS |
|---|---|---|---|

| ACTIONS | **The Basics** | | | |
|---|---|---|---|---|
| B.5.1 | Instead of buying a new item, I will try fixing, patching or retrofitting something, specifically this: _____ | | | |
| B.5.2 | Instead of buying, I will rent or borrow something I need, but will seldom use, specifically this: _____ | | | |
| B.5.3 | I will shop at garage sales, thrift stores, consignment stores, auctions, antique stores, a flea market, or search the classified section for used items to find an item that I need this month. | | | |

| ACTIONS | **I Can Do More** | | | |
|---|---|---|---|---|
| B.5.4 | I will shop through second hand internet sites such, as the Buy & Sell or Craig's List for a necessary item this month. 🌐 | | | |
| B.5.5 | Inspired by these ideas, I will take the following action perfectly suited for my lifestyle: _____ | | | |
| B.5.6 | I commit to doing action #(s) ____ consistently for ____ days/weeks/months. | | | |

| ACTIONS | **Educating and Influencing Others** | | | |
|---|---|---|---|---|
| B.5.7 | I will initiate and or participate in bartering for an item or service, specifically this: _____ 🌐 | | | |
| B.5.8 | I will share something with family and friends, specifically: _____ _____ | | | |

| ACTIONS | **Community Projects** | | | |
|---|---|---|---|---|
| B.5.9 | We will initiate or participate in an "exchange" event to swap such things as clothing, books, toys, paint, tools, or other items. | | | |
| B.5.10 | We will look into ___ pieces of workplace equipment to identify if it needs maintenance or inspection in order to maximize its life span. | | | |
| B.5.11 | We will ask a friend, coworker or family member to join us in completing action #(s) _____ from this intention. | | | |

## EXPLORATIONS 🌐

Habitats for Humanity's Re-stores sell and accept donations of used building supplies, appliances, and assorted home decor item. In BC: *www.vancouverhabitat.bc.ca.* For international information *www.habitat.org.*

# INTENTION 6 AVOID DISPOSABLES

Choosing to buy and use durable, reusable items rather than disposable items, whenever possible, is a cornerstone for a sustainable life style. Making a commitment to use durable, reusable items helps to reduce the amount of solid waste going into landfills, conserve energy, and reduce our dependency on non-renewable resources. Choosing to use durable goods that can be reused or refilled is also more cost-effective in the long run, and thereby contributes to wise money management.

| | | PLEDGE DATE | # TIMES / WEEK | # TIMES COMPLETE / 7 WEEKS |
|---|---|---|---|---|
| ACTIONS | **The Basics** | | | |
| B.6.1 | As they are needed, I will purchase reusable cloth items instead of their disposable paper equivalents, including: dish cloths___ napkins___ paper towels___ handkerchiefs___ cleaning rags___ washable mop heads___ diapers ___ other_____. | | | |
| B.6.2 | I will purchase and use a reusable mug and will save the equivalent disposable cups from entering the landfill. ✏ | | | |
| B.6.3 | I will use refillable pens and pencils. | | | |
| B.6.4 | I will list the items I am currently using that are disposable and investigate longer-lasting alternatives. | | | |
| B.6.5 | I will buy the more durable, long lasting option next time I go to purchase_____. | | | |
| ACTIONS | **I Can Do More** | | | |
| B.6.6 | I will purchase a reusable razor. | | | |
| B.6.7 | I will purchase and use ___ rechargeable batteries. | | | |
| B.6.8 | I will buy and use a reusable lunch bag. | | | |
| B.6.9 | I will purchase kitchen canisters (preferably metal or glass) so I can buy and store supplies in bulk. | | | |
| B.6.10 | Inspired by these ideas, I will take the following action perfectly suited for my lifestyle: _____ | | | |
| B.6.11 | I commit to doing action #(s) ____ consistently for ____ days/weeks/months. | | | |
| ACTIONS | **Educating and Influencing Others** | | | |
| B.6.12 | I will bring re-usable to-go containers for take-out food orders to replace the disposable paper, plastic or styrofoam that the establishment provides. | | | |
| B.6.13 | Next time I am out with a friend that is using a disposable paper cup, I will mention to them that it is a non-recyclable item because of its plastic coating. 🌍 | | | |
| B.6.14 | I will have a challenge among a group of individuals on who can go the longest without using a disposable cup or plastic bag. | | | |

# INTENTION 6  **AVOID DISPOSABLES**

| | | PLEDGE DATE | # TIMES/ WEEK | # TIMES COMPLETE / 7 WEEKS |
|---|---|---|---|---|
| B.6.15 | Next social gathering I host or go to, I will use non-disposable cutlery, napkins and dishes, and help/recruit others to help with the dishes and cleanup. | | | |
| B.6.16 | I will ask a friend, coworker or family to join me in completing action #(s) _____ from this intention. | | | |
| Actions | **Community Projects** | | | |
| B.6.17 | We will ensure that any new items in the staff lunchroom, such as plates, utensils, mugs, and glasses, come from second-hand sources. | | | |
| B.6.18 | We will encourage __, advocate__, ensure__ that our organizations minimize and eventually eliminate the use of disposable items. | | | |
| B.6.19 | We will personally implement action #(s) _____ from this intention at our schools and/or workplaces. | | | |
| B.6.20 | We will request that our workplaces and/or schools consider implementing action #(s) _____ from this intention; we will research on how this is done to make the process as easy as possible. | | | |

## EXPLORATIONS

The frightening facts about disposable paper cups: Paper cups are not made of recycled paper and are not recyclable! In 2006, it is estimated that paper cups accounted for 252 million pounds of garbage in landfills, 4 billion gallons of water wasted, 6.5 million trees cut down, and 4,884 billion BTU's of energy used. These facts and more are in an excellent summary of the environmental problem of coffee cups at: *www.sustainabilityissexy.com/facts.html*
Why your reduction in consuming disposable items will make a difference: *www.helium.com/items/1463592-easy-steps-to-reducing-consumption*

# INTENTION 7  AVOID USING PLASTICS & PACKAGING

Product packaging accounts for as much as 1/3 of the solid waste produced by households each day![4] This packaging is often not recyclable or is hard to recycle. Much of the packaging used for food, beverages and other items are plastics manufactured from non-renewable resources such as petroleum-based products. Many of these plastics often carry additional chemicals hazardous to human health and our environment.[5] Plastics require 100 to 400 years to break down in a landfill, yet every year the United States makes enough plastic film to shrink-wrap the state of Texas.[6] It's time to change this, starting with our daily actions and habits, then moving towards greater systemic change.

| | | | | |
|---|---|---|---|---|
| Actions | **The Basics** | | | |
| B.7.1 | I will use cloth shopping bags instead of plastic ones.  | | | |

4. www.headwatersrecycle.com/why.html
5. Gershon, David. Low Carbon Diet (2007), p.7
6. www.headwatersrecycle.com/why.html

# INTENTION 7 **AVOID USING PLASTICS & PACKAGING**

| | | PLEDGE DATE | # TIMES/ WEEK | # TIMES COMPLETE / 7 WEEKS |
|---|---|---|---|---|
| B.7.2 | After changing to cloth bags, I will recycle my plastic bags at stores with plastic bag recycling (Safeway, London Drugs, Save-On-Foods), or contact my local municipality or regional district to find other locations. | | | |
| B.7.3 | I will refrain from purchasing bottled water. ✏ | | | |
| B.7.4 | I will bring a stainless steel or glass water bottle with me. | | | |
| ACTIONS | **I Can Do More** | | | |
| B.7.5 | I will avoid purchasing items in packaging that cannot be recycled (blister packages, individually wrapped snacks, single-serving microwaveable containers, and polystyrene foam including egg cartons, meat trays) | | | |
| B.7.6 | I will store my food and beverages (including water) in containers not made from plastic wrap or Polyvinyl chloride (Plastic #3), Polycarbonate plastic Bisphenol A (Plastic #7) or polystyrene foam (Plastic #6). | | | |
| B.7.7 | I will take garbage-free lunches this week. A garbage-free lunch involves: no juice boxes (tetra-packs), straws, pre-packaged snacks, single-serving microwavable lunches, plastic cutlery, paper napkins or plastic bags. | | | |
| B.7.8 | I will bring my own containers and purchase goods in bulk. | | | |
| B.7.9 | I will do an in-house audit to determine ways to reduce and eliminate plastics and packaging waste. Is it possible to eliminate packaging from my waste stream? | | | |
| B.7.10 | Inspired by these ideas, I will take the following action perfectly suited for my lifestyle: _____ | | | |
| B.7.11 | I commit to doing action #(s) ____ consistently for ____ days/weeks/months. | | | |
| ACTIONS | **Educating and Influencing Others** | | | |
| B.7.12 | I will send a clear message to retailers and manufacturers by leaving packaging at the retail counter. | | | |
| B.7.13 | I will locate and frequent stores in my neighborhood that provide green-products and have environmentally-sound practices in place, and I will request that my other local stores sell these products. | | | |
| B.7.14 | I will advocate for my school or workplace to be plastic water bottle free (making sure there are water sources available, such as water fountains). | | | |
| B.7.15 | I will research into some of the products we require for work that have plastic packaging and I will find alternatives that use less packaging and bring my findings to a work decision maker. | | | |
| B.7.16 | I will ask a friend, coworker or family to join me in completing action #(s) _____ from this intention. | | | |

# INTENTION 7  AVOID USING PLASTICS & PACKAGING

| | | PLEDGE DATE | # TIMES / WEEK | # TIMES COMPLETE / 7 WEEKS |
|---|---|---|---|---|
| Actions | **Community Projects** | | | |
| B.7.17 | We will research alternatives to plastic cups and other disposable plastic products used by establishments that we frequently eat at, and give them other economical options to replace the plastic with. | | | |
| B.7.18 | We will work with our neighbours and/or co-workers to set up a plastics recycling depot where people can recycle plastics that are not accepted by their municipalities. 🌐 | | | |
| B.7.19 | We will work together to ban the supply of plastic bags at a local grocery store. | | | |
| B.7.20 | We will organize and advocate an annual trash-free day at our workplaces and/or schools, where our coworkers/peers and I aim to have no garbage from our food or purchases for an entire day. | | | |

## EXPLORATIONS 🌐

For "How To Start A Recycling Depot" see Resources on *www.bethechangeEarthAlliance.org*

Why reducing plastic consumption will make a difference:
*www.helium.com/items/1463592-easy-steps-to-reducing-consumption*

Why the biggest plastic landfill is in the Pacific Ocean:
*www.science.howstuffworks.com/great-pacific-garbage-patch.html*

Waves of Disaster: *www.straight.com/article-157369/waves-disaster?rotator=1*

Identifying plastics by number: *www.davidsuzuki.org/publications/downloads/2010/plasticsbynumber.pdf*

There are other options, example companies supplying biodegradable cups and to-go containers:
*http://newbiodegradable.com/*or *http://www.biodegradablerus.com/*

Bisphenol A (Plastic #7) and your health: *http://www.cbc.ca/news/health/story/2009/01/28/f-health-bisphenol.html*

Plastic bags and their impact on our environment:
*www.natural-environment.com/blog/2008/01/10/environmental-impact-of-plastic-bags*

How to Reduce, Reuse, and Recycle Plastic Shopping Bags:
*www.environment.about.com/od/reducingwaste/a/no_plastic_bags.htm*

## INTENTION 8 REUSE EVERYTHING

Being able to repair items and discover creative new ways to use objects, that have seen their better days, and finding ways to not buy new "stuff" are all important ways to follow through on the Reuse part of the Reduce, Reuse, Recycle. One way to navigate out of a consumer driven society into one that values longevity and resourcefulness is to make our resources last - by buying for keeps and learning how to make do with what we have.

| | | | | |
|---|---|---|---|---|
| Actions | **The Basics** | | | |
| B.8.1 | I estimate that I will repair __ things in this month. Specifically, I intend to repair: _____. | | | |

# INTENTION 8  **REUSE EVERYTHING**

| | | PLEDGE DATE | # TIMES/ WEEK | # TIMES COMPLETE / 7 WEEKS |
|---|---|---|---|---|
| B.8.2 | I will sell ___ unwanted items on-line, at auctions, on consignment, or at garage sales, etc., this month. | | | |
| B.8.3 | I will donate ___ items that I no longer want or need to charities or friends in need. | | | |
| B.8.4 | I will bring a stainless steel or glass water bottle with me. | | | |

| ACTIONS | **I Can Do More** | | | |
|---|---|---|---|---|
| B.8.5 | I will learn ___new skills this month, such as teaching myself to fix things, to become more self-reliant at basic carpentry, plumbing, appliance repair, gardening, crafts, sewing or darning. Specifically, I intend to: _____ | | | |
| B.8.6 | I will attend a DYI (Do it yourself) event or workshop, to see what else I can accomplish myself. | | | |
| B.8.7 | I will wrap ___ gifts in colorful comics, recycled gift-wrapping, old maps, leftover cloth, or other things this month. | | | |
| B.8.8 | I will donate ___ books this month to libraries, schools, or sell them to used bookstores, or exchange them through book exchanges. | | | |
| B.8.9 | I will re-gift ___ gifts that I received and do not use. | | | |
| B.8.10 | I will find one item in my house that is meant to be used for one thing, but actually can be used for one or more other uses. | | | |
| B.8.11 | I commit to doing action #(s) ____ consistently for ____ days/weeks/months. | | | |

| ACTIONS | **Educating and Influencing Others** | | | |
|---|---|---|---|---|
| B.8.12 | I will take a course in a new skill with a friend, coworker or family member. | | | |
| B.8.13 | I will volunteer to fix a friend, coworker, or family members'_____ (item) with my new or old fix-it skills. | | | |
| B.8.14 | I will ask a friend, coworker or family to join me in completing action #(s) _____ from this intention. | | | |

| ACTIONS | **Community Projects** | | | |
|---|---|---|---|---|
| B.8.15 | We will ensure that for the next ___ number of function(s) at work, such as birthday celebrations and special lunches, that we use only re-usable plates, cutlery and napkins. | | | |
| B.8.16 | We will hold a local craft fair in which people bring things they are not using, and children or adults take what they want/re-create what they want with the supplies and items brought by themselves or others. | | | |
| B.8.17 | We will help coordinate a DYI (Do It Yourself) event. | | | |

# INTENTION 9  RECYCLE EVERYTHING

Last, but certainly not least, Recycling is the third "R" in the "Reduce, Reuse, Recycle" strategy, and it plays a crucial role in reducing the overall amount of solid waste that ends up in a landfill. Most households produce roughly 2 kgs (4.5 lbs) of solid waste per day. That is equivalent to 730 kgs (1,642 lbs) per year! Considering that every kg or lb of solid waste in the landfill produces twice as much greenhouse gas, choosing to recycle as many objects as possible significantly helps reduce the amount of carbon dioxide ($CO_2$) and other greenhouse gas emissions such as methane ($CH_4$).[7] For every litre of garbage we are able to reduce, saves 0.2 kilograms of $CO_2$ emissions from being released into our environment![8]

| ACTIONS | **The Basics** | PLEDGE DATE | # TIMES/ WEEK | # TIMES COMPLETE / 7 WEEKS |
|---|---|---|---|---|
| B.9.1 | I will locate information on my local municipal or regional recycling collection depot or transfer station, and make sure I am following the correct recycling procedures. 🌎 | | | |
| B.9.2 | I will ensure that there are user-friendly recycling systems in my home or apartment this month. If one exists, I will improve upon the system and if there isn't, I will start one. 🌎 | | | |
| B.9.3 | I will take ___ items that are not suitable for landfill and cannot be reused, to my nearest recycling depot or municipal or regional transfer station, or to a scrap metal dealer. The items below that apply to my action are circled, or I've added them to this list. 🌎  <br> tires          car battery          microwave oven <br> fridge         barbecue             lawn chairs          bicycle | | | |

| ACTIONS | **I Can Do More** | | | |
|---|---|---|---|---|
| B.9.4 | I will recycle all my alkaline batteries. Record # of batteries in Table. Many stores where batteries are sold also take batteries for recycling, such as London Drugs in BC, and across Canada at Mountain Equipment Co-op. 🌎 | | | |
| B.9.5 | I will take ___ irreparable electronics to a facility that guarantees the electronics are recycled in an environmental sound manner (companies or facilities that do not recycle off-shore). 🌎 | | | |
| B.9.6 | I will do an in-house or workplace audit to determine ways to further reduce and eliminate waste. How close can I get to Zero Waste? | | | |
| B.9.7 | I will find an organization like Charity Village where I can donate things (art supplies, medical equipment, furniture, household items, clothes, electronics, building materials and vehicles) and donate the following items: _____ _____ | | | |
| B.9.8 | I will donate my old eyeglasses to Third World Eye Care (www.twecs.ca/recycle.php) or drop them off at a local optometrists or eye care professional who will donate them to Third World Eye Care. | | | |

7. Gershon, David. *Low Carbon Diet* (2007), p. 7
8. Ibid p. 8

# INTENTION 9 RECYCLE EVERYTHING

| | | PLEDGE DATE | # TIMES / WEEK | # TIMES COMPLETE / 7 WEEKS |
|---|---|---|---|---|
| B.9.9 | Inspired by these ideas, I will undertake the following action that suits me perfectly: (being as specific and measurable as possible will help me clarify and complete it) | | | |
| B.9.10 | I commit to doing action #(s) ____ consistently for ____ days/weeks/months. | | | |

| ACTIONS | **Educating and Influencing Others** | | | |
|---|---|---|---|---|
| B.9.11 | I will invite neighbours and friends to be part of the solution and have a conversation with ____ people about landfill bans, plastics, waste reduction and other topics in this section. | | | |
| B.9.12 | I will let ___ retailers know what I want and I don't want, in terms of product manufacturing, packaging and display. | | | |
| B.9.13 | I will request and host a Zero Waste Challenge workshop. | | | |
| B.9.14 | I will ask a friend, coworker or family to join me in completing action #(s) _____ from this intention. | | | |

| ACTIONS | **Community Projects** | | | |
|---|---|---|---|---|
| B.9.15 | At our workplaces, we will commit to being in charge of ___ item(s) of equipment, such as furniture, computers, computer disks, telephones, fridges, stoves, batteries, and/or other items, to ensure that they are donated or recycled at appropriate places rather than being thrown away. | | | |
| B.9.16 | We will make sure that there are cardboard recycling programs at our workplaces. If they are in place, we will make sure they are easily accessible to others. | | | |
| B.9.17 | If there are no cardboard recycling programs at our workplaces, we will research potential programs and present the results to a decision-maker. | | | |
| B.9.18 | We will make sure that there are aluminum, glass, plastics, and beverage carton recycling programs at our workplaces. If they are in place, we will make sure that they are easily accessible and organized. | | | |
| B.9.19 | If there are no aluminum, glass, plastics, and beverage carton recycling programs at our workplaces, we will research potential programs and present the results to a decision-maker. | | | |
| B.9.20 | We will check to see if our workplaces and/or schools return toner cartridges to manufacturers for recycling. If not, we will research ways to make this possible. | | | |
| B.9.21 | We will create a compost pool with our neighbours. | | | |
| B.9.22 | We will approach school boards about recycling education programs. | | | |

# INTENTION 9  **RECYCLE EVERYTHING**

| | | PLEDGE DATE | # TIMES / WEEK | # TIMES COMPLETE / 7 WEEKS |
|---|---|---|---|---|
| B.9.23 | We will engage a group of people (such as neighbours, members of community centre, fire hall, faith group, school, or youth club) to create a collection and delivery day for items that are not included in the municipal recycling system, but are not suitable for the landfill. (batteries, paints, soft plastics, hazardous chemicals, etc.)  | | | |
| B.9.24 | We will lobby government to legislate manufacture packaging (e.g. packaging must be fully recyclable; manufacturer return-to-sender programs; packaging made from local post-consumer product). | | | |
| B.9.25 | We will personally implement action #(s) _____ from this intention at our schools and/or workplaces. | | | |
| B.9.26 | We will request that our workplaces and/or schools consider implementing action #(s) _____ from this intention; we will research how this is done to make the process as easy as possible. | | | |

## EXPLORATIONS

There are probably websites specific to your municipality that can give you the information you are looking for.

For example, there is info on Recycling and the Zero Waste Challenge in Metro Vancouver at this website: *www.metrovancouver.org/services/solidwaste/recycling/Pages/default.aspx or www.charityvillage.com/cv/charityvillage/donate.asp*

For the Fraser Valley Regional District (FVRD): *www.fvrd.bc.ca/SERVICES/GARBAGEANDRECYCLING/FVRDGARBAGERECYCLINGFACILITIESSERVICES/Pages/default.aspx*

How To Go Green At Work: *www.planetgreen.discovery.com/go-green/green-work*

BC Hydro Green Your Business:

*www.bchydro.com/guides_tips/green_your_business/office_guide/Buy_sustainable_recyclable_supplies.html*

How to Green a School in British Columbia: *www.bced.gov.bc.ca/greenschools*

How to Make Money from Scrap Metal: *www.xomba.com/how_to_make_money_recycling_scrap_metal*

For recycling alkaline batteries in BC:

- The Society Promoting Environmental Conservation (SPEC): www.spec.bc.ca/greenpages/question.php?questID=71
- The Recycling Council of British Columbia: www.rcbc.bc.ca
- In the Lower Mainland of BC Call: 604-RECYCLE (604-732-9253)
- Rest of BC: 1-800-667-4321

For information on electronic waste:

- Information on electronic waste: *www.greencalgary.org/images/uploads/File/ElectronicsWaste_GC.pdf*
- Why electronic waste is harmful if recycled overseas?: *www.recycle.ubc.ca/ewaste.htm*
- Free Geek Vancouver: *www.freegeekvancouver.org/Portland Oregon, USA: www.freegeek.org*

*"When you change the way you look at things, the things you look at change."*

– Max Planc, Nobel physicist

## VALUE B SUPPLEMENT

## AVOID THE LURE OF ADVERTISING

This week I will journal and construct a hierarchy of my own basic needs those things that are absolutely necessary for me to live, to be fulfilled, and to be happy.

Refer to Abraham Maslow's Hierarchy of Needs outlined below, and list up to 10 things in each developmental stage that will fulfill your needs in that level. Then narrow it down to the top 3-5 for each level. When you move forward in your life, keep these priority needs in mind so that you do not waste time and energy on distractions, and can focus your attention on addressing your priority needs.

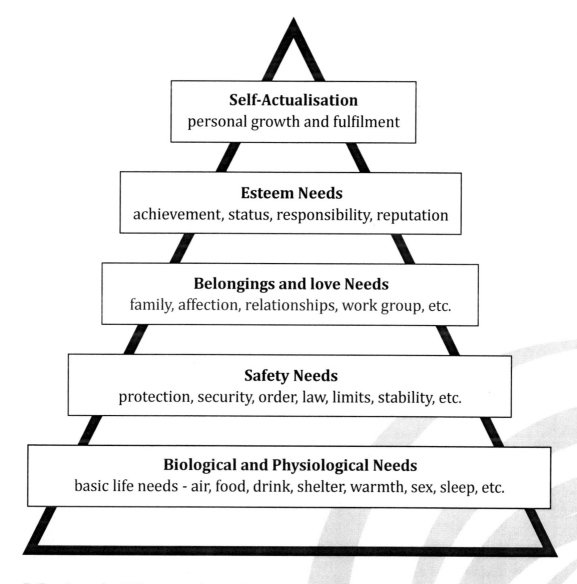

**Self-Actualisation**
personal growth and fulfilment

**Esteem Needs**
achievement, status, responsibility, reputation

**Belongings and love Needs**
family, affection, relationships, work group, etc.

**Safety Needs**
protection, security, order, law, limits, stability, etc.

**Biological and Physiological Needs**
basic life needs - air, food, drink, shelter, warmth, sex, sleep, etc.

# Maslow's Hierarchy of Needs (original five-stage model)

© alan chapman 2001-4, based on Maslow's Hierarchy of Needs

Reprinted with permission from the Alan Chapman Consultancy

# VALUE C
# CONSERVE ENERGY & NATURAL RESOURCES

*I value the limited gifts and diverse life forms of the natural world and will not waste or destroy them thoughtlessly.*

*"Human activity is putting such a strain on the natural functions of Earth that the ability of the planet's ecosystem to sustain future generations can no longer be taken for granted."[6]*

That's the call from our esteemed scientists. The human activity they're talking about is ours! With opened eyes and understanding, we can choose to be the generation that started to restore the creative diversity of our planet, rather than destroy it. We can use our commitment and creativity to conserve the energy, wilderness and natural resources that still remain so that our planet remains abundant for our children's future.

Here's the facts. The world's old growth forests are the lungs of the planet and home to endangered wildlife and indigenous peoples. Over 75% have been logged or degraded.[7] What about our water? 97.5% of the earth's water is saltwater; if the world's water fit into a bucket, only one teaspoonful would be drinkable. Our actions are seriously threatening the sanitation and availability of that 2.5% left for all the species to share.[8]

Who is the 5% of the population that is using 30% of the world's resources and creating 30% of the world's waste?[9] We are. Oil and coal combustion is the major contributor to increasing carbon dioxide in the atmosphere, thereby contributing to probable global warming. Climate change is going to have a large impact on food production and all the natural processes vital to a productive environment[10] The time to take action is now, so we can be part of a balanced and thriving world tomorrow.

## I Intend To:

| | |
|---|---|
| *C 1 | **Drive Less** |
| C 2 | **Reduce the Impact of My Driving** |
| C 3 | **Cut Back on Air Travel** |
| C 4 | **Save Paper & Save Trees** |
| *C 5 | **Use Energy Efficient Lighting** |
| C 6 | **Energy Efficient Check-Up** |
| *C 7 | **Save Water & Energy: Laundry** |
| *C 8 | **Save Water & Energy: Bathing** |
| C 9 | **Save Water & Energy: Kitchen & Yard** |
| *C 10 | **Furnace & Water Heater Efficiency** |
| *C 11 | **Cool Down More Efficiently** |
| C 12 | **Switch to Renewable Energy** |
| C 13 | **Offset My Carbon Footprint** |

6. Millennium Ecosystem Assessment Report from 1300 scientists from around the world: www.millenniumassessment.org/en/Index.aspx
7. www.csiwhalesalive.org/ny4whales/treefree.html
8. www.wateraidamerica.org/what_we_do/statistics.aspx?gclid=COy0ocHRgaoCFSN5gwodGnnO0g#top
9. Annie Leonard, The Story of Stuff: www.storyofstuff.com
10. http://dieoff.org/page84.htm

# *INTENTION 1  DRIVE LESS

Driving is an unsustainable practice that adds significantly both to greenhouse gas emissions, and air pollution that contributes to respiratory ailments. Driving plays a major part in global climate change at the individual and national levels. Since every litre of gasoline used produces 2.4 kg of $CO_2$,[1] opting to drive less often or reducing the number of single occupant vehicles on our roads makes a big difference to our environment, happiness and pocketbooks!

| ACTIONS | The Basics ✐ | PLEDGE DATE | # TIMES/ WEEK | # TIMES COMPLETE / 7 WEEKS |
|---|---|---|---|---|
| C.1.1 | I will coordinate several errands in one trip. | | | |
| C.1.2 | I will use public transit instead of driving. ⊕ | | | |
| C.1.3 | I will walk or bike to closer destinations. | | | |
| C.1.4 | I will schedule and practice a 'no-car day'. | | | |

| ACTIONS | I Can Do More ✐ | | | |
|---|---|---|---|---|
| C.1.5 | I will carpool with families for children's activities. | | | |
| C.1.6 | I will organize to carpool with others to work. | | | |
| C.1.7 | I will try out a ride-share. ⊕ | | | |
| C.1.8 | I will try out an auto-share. ⊕ | | | |
| C.1.9 | I will use email and teleconferencing for meetings to avoid driving. ⊕ | | | |
| C.1.10 | I will spend ___ hours informing myself about the environmental consequences of local highway expansion projects, including how these projects affect local wildlife, air and water quality, climate change and local trade. | | | |
| C.1.11 | I will work from home ___ times this week or I will talk to a decision-maker about working from home. | | | |
| C.1.12 | I will look into purchasing an electric bicycle. | | | |
| C.1.13 | I will keep one less car in my household. | | | |
| C.1.14 | I will investigate the option of moving closer to work, or working closer to home. | | | |
| C.1.15 | I commit to doing action #(s) ___ consistently for ___ days/weeks/months. | | | |

| ACTIONS | Educating and Influencing Others | | | |
|---|---|---|---|---|
| C.1.16 | I will post bus/train time-tables and/or other transit information in the common areas at work. | | | |
| C.1.17 | I will have a discussion with family, friends, coworkers or neighbors about how good it feels to drive less, or the positive effects of driving less. | | | |

1. http://oee.nrcan.gc.ca/transportation/tools/fuelratings/fuel-consumption-guide-2009.pdf

# INTENTION 1  **DRIVE LESS**

| | | PLEDGE DATE | # TIMES / WEEK | # TIMES COMPLETE / 7 WEEKS |
|---|---|---|---|---|
| C.1.18 | I will participate in a critical mass. 🌐 | | | |
| C.1.19 | I will research the benefits of using telephone and video conferencing as alternatives to travel for meetings and events. I will post my findings in a common area and/or I will bring it to a decision-maker at work. | | | |
| C.1.20 | I will learn about and support eco-density developments, and talk to others such as my neighbors, household, work or friends about what I have learned. | | | |
| C.1.21 | I will encourage grocers to have delivery options, so fewer cars are needed for picking up groceries. | | | |
| C.1.22 | I will reconsider my personal entitlement to owning a car. We used to feel entitled to smoke in public buildings, and now we don't. Why do we feel entitled to drive a car that pollutes everyone's air quality and causes environmental and social degradation? I will have a conversation with other people this week. | | | |
| C.1.23 | I will ask a friend, coworker or family to join me in completing action #(s)_____ from this intention. | | | |

| | | PLEDGE DATE | # TIMES / WEEK | # TIMES COMPLETE / 7 WEEKS |
|---|---|---|---|---|
| ACTIONS | **Community Projects** | | | |
| C.1.24 | We will coordinate a block-wide or neighborhood-wide ride share. | | | |
| C.1.25 | We will coordinate a block-wide or neighborhood-wide auto share. | | | |
| C.1.26 | We will write ___ letters to our local, regional and national government representatives to invest in public transit infrastructure. We will do this in collaboration with an organization, if possible. | | | |
| C.1.27 | We will research workplace or institutional public transit pass programs and bring that information to a decision maker at work. (Check with the local transit program website for information.) | | | |
| C.1.28 | We will look into the provision of facilities for bicyclists (racks, lock-up areas, showers changing rooms, etc.) in our building and we will share that information with others in the building. | | | |
| C.1.29 | We will do something to influence urban planning so people don't NEED a car to get where they need to go. Example: Calling or emailing the municipality's city planners to consider including car-free squares, meeting places and other more pedestrian-friendly areas, bike lanes and improved transit, rather than building more roads. | | | |
| C.1.30 | We will actively support existing campaigns to halt highway expansion projects. These campaigns can bring awareness to the areas that will suffer most, and inform the public about the direct effects of increased traffic and trade caused by the ensuing unsustainable development. Some ways of support include: to sign a petition, donate money to the organization, volunteer to participate in their campaign in some way. We choose to: _____ this week. | | | |

## INTENTION 1 **DRIVE LESS**

### EXPLORATIONS 🌍

Learn more about driving smart at the following:

For more suggestions on how to shrink your travel distance in your daily life, visit: *www.davidsuzuki.org/NatureChallenge/What_is_it/Transportation/*

Check your local transit website for bus routes: *www.google.com/intl/en/landing/transit/#mdy*

Vancouver Critical Mass: *http://vancouvercm.blogspot.com/*

For info on ride-sharing check out: *www.ride-share.com*

For info on auto-network: *www.modo.coop*

For more information on teleconferencing, read through the actions suggested by the WWF: *http://community.wwf.ca/ActionDetail.cfm?ActionId=39*

Or see "Getting Started," "7 Ways To Get Around," and "It's Your Business" from *http://www.gogreen.com/choices*

For information on highway expansion in BC: *http://wildernesscommittee.org/top_ten_gateway_myths*

Sustainable workplace transportation: Amenities and programs that encourage active transportation in all seasons, Transport Canada: *http://www.tc.gc.ca/eng/programs/environment-utsp-allseasontransportation-222.htm*

## INTENTION 2 REDUCE THE IMPACT OF MY DRIVING

Mixed use transportation is often appropriate, so it's great to look for ways to reduce our impact when driving. Did you know that ten seconds of idling uses more gas than restarting the engine? Or that keeping your car as aerodynamic, light and tuned up as possible can reduce your $CO_2$ output by up to 1,500 pounds this year alone?[2] The actions outlined below will cut fuel costs by 10-30% and reduce the amount of $CO_2$ in the atmosphere. As an added bonus, they can reduce stress levels and help you become a safer driver!

| ACTIONS | The Basics | PLEDGE DATE | # TIMES/ WEEK | # TIMES COMPLETE / 7 WEEKS |
|---|---|---|---|---|
| C.2.1 | I will be very aware to not have my car idle needlessly. For example, I will turn my vehicle off at train crossings, schools, a drive-through, community centres, shopping centres, gas stations and border line-ups. Ten seconds of idling uses more gas than restarting the engine. | | | |
| C.2.2 | I will drive within the speed limit. | | | |
| C.2.3 | I will drive calmly; refrain from jack-rabbit starts, and braking hard. 🌍 | | | |
| C.2.4 | I will open the window instead of using the air conditioner. | | | |
| C.2.5 | I will verify this week, and on a monthly basis thereafter that my tires are adequately inflated. | | | |

2. Gershon, David. *Low Carbon Diet* (2007), p. 30

# INTENTION 2   REDUCE THE IMPACT OF MY DRIVING

| Actions | I Can Do More | PLEDGE DATE | # TIMES/ WEEK | # TIMES COMPLETE / 7 WEEKS |
|---|---|---|---|---|
| C.2.6 | I will travel during off-peak hours to reduce idling. | | | |
| C.2.7 | I will try out new routes that may require less idling. | | | |
| C.2.8 | I will use cruise control and overdrive gears when appropriate. | | | |
| C.2.9 | I will remove the unnecessary items in my vehicle, especially the heavy ones. | | | |
| C.2.10 | I will keep my car as aerodynamic as possible out on the open highway (windows closed, empty bicycle and ski racks removed, car washed). | | | |
| C.2.11 | I will improve my car's fuel efficiency by maintaining an efficient car. I will have my car tuned up this week to check that the air filter and oil, in particular, are clean. A tuned up car can improve fuel efficiency up to 30% | | | |
| C.2.12 | I will keep track of my gas mileage and have my car serviced if there is a sudden drop. | | | |
| C.2.13 | I will make sure my families vehicles are tuned up and running efficiently, including the tire's inflated, air filter and oil clean. | | | |
| C.2.14 | I will consider buying a hybrid the next time I am in need of a new car.  | | | |
| C.2.15 | I commit to doing action #(s) ____ consistently for ____ days/weeks/months. | | | |

| Actions | Educating and Influencing Others | | | |
|---|---|---|---|---|
| C.2.16 | I will tell family, friends, coworkers or neighbors about the benefits of reduced idling | | | |
| C.2.17 | I will tell family, friends, coworkers or neighbors about the actions I have taken for this Intention. | | | |
| C.2.18 | I will ask a friend, coworker or family to join me in completing action #(s)_____ from this intention. | | | |

## EXPLORATIONS

Learn more about driving smart at the following:

http://www.ec.gc.ca/p2/default.asp?lang=En&n=F6F26D4F-1

www.davidsuzuki.org/NatureChallenge/newsletters/Three.asp

www.eartheasy.com/live_fuel_efficient_driving.htm

Hybrids and electric car information: www.davidsuzuki.org/NatureChallenge/What_is_it/Transportation/

For detailed information about vehicle idling, visit the BC anti-idling campaign: www.idlefreebc.ca/

## INTENTION 3  CUT BACK ON AIR TRAVEL

Air travel has a much bigger climate impact, per traveler, than any other mode of transportation. A return flight of approximately 1500 km (or 3,000 miles) can produce 1-2 tonnes of carbon dioxide emissions per person.[3] If I fly a lot, I am probably well above the national average of carbon emissions per individual. At present, there are no climate-friendly alternatives to the kerosene used for jet fuel, and the climate impacts of air travel are not sufficiently regulated either by national or international laws,[4] so it is very important I take personal responsibility for reducing air travel.

| | | PLEDGE DATE | # TIMES/ WEEK | # TIMES COMPLETE / 7 WEEKS |
|---|---|---|---|---|
| ACTIONS | **The Basics** | | | /YEAR |
| C.3.1 | I will take ___ less flights this year to cut back on air travel for personal travel. I will consider other places closer to home I like to vacation. | | | |
| ACTIONS | **I Can Do More** | | | /YEAR |
| C.3.2 | I will use a carbon-offset program to offset the impact of ___ flights I travel this year. | | | |
| C.3.3 | I will take ___ less flights this year to cut back on air travel for business. Research ways you can reduce your travel for work. | | | |
| ACTIONS | **Educating and Influencing Others** | | | |
| C.3.4 | I will request that a group or family trip be local, rather than flying to a destination. | | | |
| ACTIONS | **Community Projects** | | # PPL | Accomplishments |
| C.3.5 | We will advocate using teleconferencing or similar technology, and will suggest we bring speakers to our work, rather than flying employees to conferences or educational events. | | | |

## EXPLORATIONS 🌍

To discover why air travel is so harmful to the climate read:

*www.davidsuzuki.org/Climate_Change/What_You_Can_Do/air_travel.asp*

For further information: *www.terrapass.com/carbon-footprint-calculator/#air*

*"Progress is impossible without change, and those who cannot change their minds cannot change anything."* – George Bernard Shaw

3. *The Greenpeace Green Living Guide* (2007), p. 137
4. www.davidsuzuki.org/Climate_Change

# INTENTION 4 SAVE PAPER & SAVE TREES

It is easy to become inundated with all the newspapers, books, magazines, bills, flyers, junk mail and cardboard packaging that come into our homes. This is directly related to the loss of our old growth forests and the increase in global consumption of paper.[5] Most white papers are created through a chlorine bleaching process that releases toxic chemicals into rivers and oceans. The following actions will help me to save trees.

| | | PLEDGE DATE | # TIMES/ WEEK | # TIMES COMPLETE / 7 WEEKS |
|---|---|---|---|---|
| ACTIONS | **The Basics** | | | |
| C.4.1 | I will recycle all paper through my municipal system or regional transfer site this month. | | | |
| C.4.2 | I will write on both sides of my note books and pads. | | | |
| C.4.3 | I will only print out what is totally necessary, if I only need a few facts or directions from a page on the computer than I will jot it down on my recycled note book rather than printing the entire page(s). | | | |
| C.4.4 | I will practice printing on both sides of paper. | | | |
| C.4.5 | I will set my printer to automatically print double sided. | | | |
| C.4.6 | I will purchase recycled paper goods including computer paper and toilet and tissue products that have the highest content of Post Consumer Waste (PCW). 🌎 | | | |
| C.4.7 | I will choose products labeled Processed Chlorine Free (PCF). 🌎 | | | |
| C.4.8 | I will post a No Junk Mail sign or sticker this week at my doorstop, on or next to my mailbox, or if I live in an apartment on the inside of my mailbox, to notify mail or newspaper delivery persons that I no longer wish to receive unnecessary flyers, magazines, or advertisement newspapers. 🌎 | | | |
| ACTIONS | **Educating and Influencing Others** | | | |
| C.4.9 | I will put up a sign at the copier reminding others about paper conservation. | | | |
| C.4.10 | I will add "Please consider the environment before printing this email" on my email signature. | | | |
| C.4.11 | I will check to see if our paper source has a recycled content of 80% or higher. If not, I will research an alternative option and bring that to a decision-maker. | | | |
| C.4.12 | I will gather paper that has only been used on one side and create note-books or note-pads to give to family, friends or neighbors as gifts (Feel free to get creative with this). | | | |
| C.4.13 | I will start a magazine or newspaper share among my neighbors, friends, family or coworkers. | | | |
| C.4.14 | I will ask a friend, coworker or family to join me in completing action #(s) _____ from this intention. | | | |

5. The Greenpeace *Guide to Living* (2007), p.96

# INTENTION 4  **SAVE PAPER & SAVE TREES**

| Actions | Community Projects | PLEDGE DATE | # TIMES/ WEEK | # TIMES COMPLETE / 7 WEEKS |
|---|---|---|---|---|
| C.4.15 | I will request to re-set my workplace computers to default to printing on both sides. | | | |
| C.4.16 | I will explore the possibility and set up if possible, a tray in the photocopier(s) and printer(s) for one-sided recycled paper, at my workplace. | | | |
| C.4.17 | I will label a box "one-sided recycled" at high-traffic paper-use areas in my workplace. People can use the side with no printing on it in their manual feed tray for recycled printing. | | | |
| C.4.18 | I will put a recycling bin in the bathrooms for used paper towel and toilet paper roll disposal. I will make sure that the bin is emptied into the recycling program as needed. | | | |
| C.4.19 | I will go around my workplace or building to ensure that there are recycling containers at desks, at high-traffic paper-use areas, and any other areas where paper is used. If there are any recycling containers missing, I will take responsibility for supplying them. | | | |
| C.4.20 | I will look into the possibility of using hand towels in my workplace bathroom(s) rather than paper towels. | | | |
| C.4.21 | I can personally implement action #(s) _____ from this intention at my school or workplace. | | | |
| C.4.22 | I will request for my workplace or school to consider implementing action #(s) _____ from this intention; I will research on how this is done to make it as easy as possible for my company, organization or institute. | | | |

## EXPLORATIONS

To discover why air travel is so harmful to the climate read:

*www.davidsuzuki.org/Climate_Change/What_You_Can_Do/air_travel.asp*

For further information: *www.terrapass.com/carbon-footprint-calculator/#air*

Forest Stewardship Counsel of Canada (FSC) Certified Wood Products:

*www.fsccanada.org/FindWoodProducts.htm in the USA www.fscus.org*

For more Build Smart Sustainable Building Solutions: *www.metrovancouver.org/BuildSmart/Pages/default.aspxand*

What is Post Consumer Waste (PCW): *www.treecycle.com/recycling.html*

Chlorine Free Products Association: *www.chlorinefreeproducts.org /Reach For Unbleached: www.rfu.org/*

Junk Mail Stickers and Junk Mail Information: Canadian / American Red Dot Campaign at *www.reddotcampaign.ca or* make your own sticker!

USA: Remove your address from junk mail lists: *www.directmail.com/directory/mail_preference/*

*"If you think you're too small to be effective,*

*than you've never been in bed with a mosquito."* – Chinese Proverb

# *INTENTION 5  USE ENERGY EFFICIENT LIGHTING

Switching to compact fluorescent light bulbs from incandescent bulbs[6] can save me both energy and money. Compact fluorescents use 75% less energy and last 10 times longer than incandescent bulbs, so although a compact fluorescent initially costs more than an incandescent bulb, it will yield greater savings to both the environment and my pocket-book in the long run. It's important to properly recycle light bulbs of any kind.

| ACTIONS | The Basics | PLEDGE DATE | # TIMES/ WEEK | # TIMES COMPLETE / 7 WEEKS |
|---|---|---|---|---|
| C.5.1 | I will turn off the lights as I leave the room. 🖉 | | | |
| C.5.2 | I will replace the incandescent light bulbs in my home as they burn out with compact fluorescent ones. For each bulb I replace, I will save 100 pounds of $CO_2$ a year. 🖉 | | | |
| C.5.3 | I will install a power bar for my entertainment electronics and turn it off when not in use. (If I do this 7 days in a row, I'll create a positive habit so I will do this without having to think about it in the future). | | | |

| ACTIONS | I Can Do More | | | |
|---|---|---|---|---|
| C.5.4 | I will turn off the lights as I leave a room for a full day. | | | |
| C.5.5 | I will replace the incandescent light bulbs in my home as they burn out with compact fluorescent ones. For each bulb I replace, I will save 100 pounds of $CO_2$ a year. [7] | | | |
| C.5.6 | I will install a power bar for my entertainment electronics and turn it off when not in use. (If I do this 7 days in a row, I'll create a positive habit so I will do this without having to think about it in the future). | | | |
| C.5.7 | I will install motion detectors on my outside lights. | | | |
| C.5.8 | I will install energy-saving devices like dimmers, timers, and photocells. | | | |
| C.5.9 | I will make changes by noting where natural lighting may be used, rather than electric lighting, as appropriate. | | | |
| C.5.10 | I will switch to LED holiday lights. | | | |

| ACTIONS | Educating and Influencing Others | | | |
|---|---|---|---|---|
| C.5.11 | I will make little reminder signs by the light switches in my home, at my work, or school, to remind people to turn the light off after them or at the end of the day. | | | |
| C.5.12 | I will look into our current policy at my workplace for over-night and exterior lighting and I will talk to a decision-maker about making this more energy efficient if needed. | | | |

6. *The Greenpeace Green Living Guide* (2007), p. 65
7. Gershon, David. *Low Carbon Diet* (2007), p. 65

## INTENTION 5  USE ENERGY EFFICIENT LIGHTING

| | | PLEDGE DATE | # TIMES/ WEEK | # TIMES COMPLETE / 7 WEEKS |
|---|---|---|---|---|
| C.5.13 | I will challenge an organization or company to switch to more energy efficient lighting. | | | |
| C.5.14 | I will ask a friend, coworker or family to join me in completing action #(s)_____ from this intention. | | | |
| Actions | **Community Projects** | | | |
| C.5.15 | We will personally implement action #(s)_____ from this intention at our school or workplace. | | | |
| C.5.16 | We will request for our workplace or school to consider implementing action #(s) _____ from this Intention; we will research on how this is done to make it as easy as possible for our company, organization or institute. | | | |

## EXPLORATIONS

Contact your municipality to find out how to safely dispose of compact fluorescent light bulbs.

Great tips on energy efficient lighting and prices:
*www.bchydro.com/guides_tips/green-your-home/lighting_guide.html*

## INTENTION 6  ENERGY EFFICIENCY CHECK-UP

North Americans produce approximately 5 tonnes of $CO_2$ per year, per person.[8] Each person is producing about 2.5 tonnes of $CO_2$ emissions as a result of hot water tanks, heating or cooling a home, the electrical appliances and light bulbs used in the home. Between cars and homes, North Americans' individual emissions account for 28% of $CO_2$ emissions produced, overall. We can cut down on emissions produced within the home by taking some, or preferably all, of the following actions.

| | | | | |
|---|---|---|---|---|
| Actions | **The Basics** | | | |
| C.6.1 | I will install power bars and disconnect my electronics when they are not in use (TV, stereo, computer). | | | |
| C.6.2 | I will learn about energy efficiency by checking out BC Hydro's Power Smart Program: http://www.bchydro.com/powersmart/ | | | |
| C.6.3 | I will put large containers of water at the bottom of my deep freezer, if it is not filled to capacity, to reduce its energy demands. | | | |
| Actions | **I Can Do More** | | | |
| C.6.4 | I will review websites that provide information on energy audits (www.oee.nrcan.gc.ca) and conduct an on-site energy audit of my home. | | | |

8. *The Greenpeace Green Living Guide* (2007), p. 62

# INTENTION 6  **ENERGY EFFICIENCY CHECK-UP**

| | | PLEDGE DATE | # TIMES/ WEEK | # TIMES COMPLETE / 7 WEEKS |
|---|---|---|---|---|
| C.6.5 | I will upgrade my major appliances if their annual energy use is significantly higher than that of the latest energy efficient models (read the Energy Guide sticker). | | | |
| C.6.6 | If my main energy source is electricity, I will investigate switching to gas appliances or going solar. | | | |
| C.6.7 | I will replace my old fridge with a new Energy Star model. I will save 500 pounds of CO2 a year. [9] | | | |
| C.6.8 | When it's time to move, I will choose a home that is energy-efficient and meets R-2000 standards. | | | |
| ACTIONS | **Educating and Influencing Others** | | | |
| C.6.9 | I will put up a sign in high-traffic areas at work, advising others to install power bars and disconnect electronics when they are not in use. | | | |
| C.6.10 | I will put up a sign at elevators or stairwells or common-room(s) encouraging others to use the stairs for fitness and to save energy. | | | |
| C.6.11 | I will talk to my family, friends, coworkers or neighbors about the environmental and economic benefits of having a more energy efficient household, and what someone can do to achieve this. | | | |
| C.6.12 | If it is time for my workplace to move locations, I will recommend looking into buildings that are energy efficient and meets R-2000 standards. | | | |
| C.6.13 | I will ask a friend, coworker or family to join me in completing action #(s)_____ from this intention. | | | |
| ACTIONS | **Community Projects** | | | |
| C.6.14 | We will research ___ item(s) of equipment/appliances at work to see if it/they are energy efficient. | | | |
| C.6.15 | If we find inefficient equipment/appliances, or if it is time to replace old equipment/ appliances, we will research ENERGY STAR® labeling at http://oee.nrcan.gc.ca/residential/energystar-portal.cfm. We will bring this information to a decision-maker at work. | | | |
| C.6.16 | We will review websites that provide information on energy audits (www.oee.nrcan.gc.ca) and conduct an on-site energy audit of our work, and do what we can to make our work place more energy efficient. | | | |
| C.6.17 | We will review websites that provide information on energy audits (www.oee.nrcan.gc.ca) and conduct an on-site energy audit of a local organization or company and do what we can to influence them to make their space more energy efficient. | | | |

9. Gershon, David. *Low Carbon Diet* (2007), p.62

|  |  | PLEDGE DATE | # TIMES/ WEEK | # TIMES COMPLETE / 7 WEEKS |

## INTENTION 6 ENERGY EFFICIENCY CHECK-UP

### EXPLORATIONS

Learn more about energy-efficient homes and appliances at the following websites:

www.bchydro.com/guides_tips/green-your-home.html

www.bchydro.com/guides_tips/green-your-home/whole_home_efficiency.html

www.energyefficienthomearticles.com

www.davidsuzuki.org/Climate_Change/Energy/Renewables

Green your business: www.bchydro.com/guides_tips/green_your_business.htm

## *INTENTION 7 SAVE WATER & ENERGY: LAUNDRY

Electric clothes washers and dryers use a lot of energy: they generate 5 lbs (roughly 2 kilograms) of $CO_2$ per washer and dryer cycle.[10] Much of the energy associated with machine-washed clothes is spent on heating the water (90% for clothes washed in hot water) compared to actually powering the machine. Incorporating the following actions into my weekly laundry routine will reduce the amount of energy used on washing and drying clothes.

| ACTIONS | The Basics ✎ |  |  |  |
|---|---|---|---|---|
| C.7.1 | I will avoid washing clothes unless they are actually dirty, and touch up creases, spot wash, and air clothing when possible instead. |  |  |  |
| C.7.2 | I will wash my laundry in cold water instead of hot. *If I switch just one load of laundry a week to cold water, I save 100 pounds of CO2 a year.* [11] |  |  |  |
| C.7.3 | I will clean the lint filter in the dryer after every load. |  |  |  |
| C.7.4 | I will match the water level to the load size and spin the clothes well to cut down on drying time. |  |  |  |
| C.7.5 | I will dry like-weight items together. |  |  |  |

| ACTIONS | I Can Do More ✎ |  |  |  |
|---|---|---|---|---|
| C.7.6 | I will dry full loads only and place a large dry beach towel in the dryer with my wet clothes to speed the drying process. |  |  |  |
| C.7.7 | I will air-dry my clothes on a clothesline. If I hang-dry my clothes once a week I will save 260 pounds of CO2 a year. [12] |  |  |  |
| C.7.8 | I will buy an energy-efficient, front-loading washing machine the next time I need to replace my old one. I will save 500 pounds of CO2 a year. [13] |  |  |  |
| C.7.9 | I commit to doing action #(s) ____ consistently for ____ days/weeks/months. |  |  |  |

10. Gershon, David. (2007), p. 11
11. Ibid p. 11
12. Ibid p. 12
13. Ibid p. 11

## INTENTION 7  SAVE WATER & ENERGY: LAUNDRY

| | | PLEDGE DATE | # TIMES/ WEEK | # TIMES COMPLETE / 7 WEEKS |
|---|---|---|---|---|
| ACTIONS | **Educating an Influencing Others** | | | |
| C.7.10 | I will ask a friend, coworker or family to join me in completing action #(s) _____ from this intention. | | | |

## EXPLORATIONS 🌍

Info on Rebate programs for appliances:

*http://www.bchydro.com/rebates_savings/appliance_rebates/appliance_rebate.html*

For washing in cold water:

*http://www.bchydro.com/guides_tips/green-your-home/appliances_guide/washing_laundry.html*

For shrinking your dryers energy:

*http://www.bchydro.com/guides_tips/green-your-home/appliances_guide/drying_laundry.html*

## *INTENTION 8  SAVE WATER & ENERGY: BATHING

Many of us luxuriate in taking a long, hot shower or bath to relax while we clean on a regular basis, but how much energy are we using to run that hot water—and how much water is literally going down the drain? The energy required to heat the water for a 10 minute shower can produce as much as 4 lbs (1.8 kgs) of $CO_2$ emissions,[14] so if a person takes a 10 minute shower every day, that quickly adds up to 1,460 lbs (657 kgs) a year in $CO_2$ emissions! A 10 minute shower also uses 53 gallons (200 litres) of water.[15] On the topic of wasting water, older models of toilets use up to 5.2 gallons (20 litres) of water per flush.[16] The following actions will help reduce $CO_2$ emissions and water usage in the washroom.

| | | PLEDGE DATE | # TIMES/ WEEK | # TIMES COMPLETE / 7 WEEKS |
|---|---|---|---|---|
| ACTIONS | **The Basics** | | | |
| C.8.1 | I will turn off the tap when I brush my teeth or shave. | | | |
| C.8.2 | I will reduce my shower time by 2 minutes per shower. ✏ | | | |
| C.8.3 | I will reduce my shower time to a total maximum of 5 minutes. ✏ | | | |
| C.8.4 | I will save energy by towel drying my hair. | | | |

| | | PLEDGE DATE | # TIMES/ WEEK | # TIMES COMPLETE / 7 WEEKS |
|---|---|---|---|---|
| ACTIONS | **I Can Do More** | | | |
| C.8.5 | If the capacity of my toilet tank is more than 12 litres, I will put a sand-filled container inside my toilet tank to reduce tank volume. | | | |

14. Gershon, David. *Low Carbon Diet* (2007), p. 9
15. Based on 2007 data from Environment Canada, The World Vision's "Hard Facts" web page notes that Canadians use 100 litres of water for a five minute shower. www.worldvision.ca/EDUCATION-AND-JUSTICE/ADVOCACY-IN-ACTION/Pages/the-hard-facts.aspx
16. *The Greenpeace Green Living Guide* (2007), pp. 149-150

# INTENTION 8  SAVE WATER & ENERGY: BATHING

| | | PLEDGE DATE | # TIMES/ WEEK | # TIMES COMPLETE / 7 WEEKS |
|---|---|---|---|---|
| C.8.6 | I will practice minimal flushing. (If family members follow the 'yellow mellow' even once between toilet uses, it will significantly reduce my household's daily water consumption.) | | | |
| C.8.7 | I will use a bucket to catch warm-up water from the shower and use it to flush the toilet. | | | |
| C.8.8 | I will install low-flow showerheads. ✏ | | | |
| C.8.9 | I will purchase a low-flush or dual flush toilet. 🌐 | | | |
| C.8.10 | I commit to doing action #(s) ____ consistently for ____ days/weeks/months. | | | |

| ACTIONS | **Educating or Influencing Others** | | | |
|---|---|---|---|---|
| C.8.11 | I will ask my friends, family or neighbors how they reduce their water consumption, and in turn explain to them what I have done. | | | |
| C.8.12 | I will ask a friend, coworker or family to join me in completing action #(s)_____ from this intention. | | | |

| ACTIONS | **Community Projects** | | | |
|---|---|---|---|---|
| C.8.13 | We will research into how our work could lower its water consumption and take steps towards implementing it. | | | |
| C.8.14 | We will research into how a company or organization could lower its water consumption and take steps towards implementing it. | | | |
| C.8.15 | We will personally implement action #(s)_____ from this intention at our school or workplace. | | | |
| C.8.16 | We will request for our workplace or school to consider implementing action #(s) _____ from this intention; we will research on how this is done to make it as easy as possible for our company, organization or institute. | | | |

# EXPLORATIONS

www.davidsuzuki.org/NatureChallenge/newsletters/nov2004_greendesign/page6.asp

www.bchydro.com/guides_tips/green-your-home/water_guide.html

For saving water in your bathroom:

http://www.bchydro.com/guides_tips/green-your-home/water_guide/Save_water_in_the_bathroom.html

For whole home efficiency, talks about low flow toilets, and water efficiency as a whole:

http://www.bchydro.com/guides_tips/buyersguide/Whole_Home_Efficiency.html

*"We become that which we repeatedly do.*

*Excellence, then, is not a single act but a habit."* – Aristotle

# INTENTION 9 SAVE WATER & ENERGY: KITCHEN & YARD

Washing the dishes generates between 2-3 lbs (0.9-1.4 kgs) of $CO_2$ and may use up to 15 gallons of water,[17] depending on whether we use a dishwasher or wash dishes by hand. In addition to reducing the amount of $CO_2$ we emit, we also need to be mindful of the amount of water we use for washing, cleaning and watering lawns—particularly as, on average, each North American household uses approximately 500,000 litres of water per year.[18] To cut down on energy and water consumption, plan to implement the strategies listed below to start reducing both $CO_2$ emissions and water usage in your home.

| Actions | The Basics | PLEDGE DATE | # TIMES/ WEEK | # TIMES COMPLETE / 7 WEEKS |
|---|---|---|---|---|
| C.9.1 | I will wait until the dishwasher is full before turning it on. By reducing dishwasher use by one load a week, I will save 100 pounds of $CO_2$ a year. [19] | | | |
| C.9.2 | When boiling water, I will keep the lid on and reduce the heat as soon as it has boiled. | | | |
| C.9.3 | I will not overfill the kettle. | | | |
| C.9.4 | I will hand wash dishes by filling a basin with water to wash, and another to rinse. By minimizing my hot water use and hand washing my dishes, I will save approximately 125 pounds of $CO_2$ a year. [20] | | | |
| C.9.5 | I will be prudent with how much water I use to wash the car. | | | |
| C.9.6 | I will set my dishwasher to its energy efficient mode and use "air dry" not "heat dry". | | | |
| C.9.7 | I will water my garden early in the morning or in the evening to reduce the amount of water the plants require. | | | |
| C.9.8 | I will keep the grass longer through the summer so it requires less water. | | | |

| Actions | I Can Do More | | | |
|---|---|---|---|---|
| C.9.9 | I will research into efficient irrigations systems for my garden. | | | |
| C.9.10 | I will research ways in which "graywater" can be used in and outside of my house. | | | |
| C.9.11 | I will use the rinse water to water my indoor and patio plants. | | | |
| C.9.12 | I will setup rain barrels. | | | |
| C.9.13 | I will install low-flow aerators on ___ of my faucets and stop all leaks. | | | |
| C.9.14 | I will turn my yard into a xeriscape garden that requires no extra watering. For ideas visit www.eartheasy.com/grow_xeriscape.html | | | |
| C.9.15 | I will use a thermometer to make sure that the refrigerator(s) in my home is set to about 40 Fahrenheit / 4 degrees Celsius; and that the freezer is set to about 14 degrees Fahrenheit / -10 degrees Celsius. | | | |

17. Gershon, David. *Low Carbon Diet* (2007), p.10
18. *The Greenpeace Green Living Guide* (2007), p. 146
19. Gershon, David. *Low Carbon Diet* (2007), p.10
20. Ibid p. 10

# INTENTION 9 **SAVE WATER & ENERGY: KITCHEN & YARD**

| | | PLEDGE DATE | # TIMES / WEEK | # TIMES COMPLETE / 7 WEEKS |
|---|---|---|---|---|
| C.9.16 | I will clean the gasket and sealing surfaces of the refrigerator(s) and freezer(s) to ensure that warm air is not getting in my home. | | | |
| C.9.17 | I will vacuum the coils behind and below the refrigerator(s) in my home. This will allow the compressor to work more smoothly, using less energy. | | | |
| C.9.18 | I will buy an energy efficient dishwasher when it is time to replace my old one. By doing this I will save 100 pounds of CO2 a year.[21] | | | |
| C.9.19 | I commit to doing action #(s) _____ consistently for _____ days/weeks/months. | | | |
| ACTIONS | **Educating and Influencing Others** | | | |
| C.9.20 | I will start a dialogue about use of water and ways to be efficient with my friends, family, neighbors or stranger. | | | |
| C.9.21 | I will research local plumbing companies that use sustainable practices. I will bring my research to a decision-maker at work. | | | |
| C.9.22 | If applicable, I will research companies that use sustainable landscaping practices, such as water-wise landscaping, native-plant species gardening, use of grey-water/rain-water barrels, and/or pesticide-free products. I will recommend this company to a decision-maker at work. | | | |
| C.9.23 | I will ask a friend, coworker or family member to join me in completing action #(s) _____ from this intention. | | | |
| ACTIONS | **Community Projects** | | | |
| C.9.24 | We can personally implement action #(s)_____ from this intention at our school or workplace. | | | |
| C.9.25 | We will request for our workplace or school to consider implementing action #(s) _____ from this intention; we will research on how this is done to make it as easy as possible for our company, organization or institute. | | | |

## EXPLORATIONS 🌍

Research other ways to conserve water at home:
*www.davidsuzuki.org/NatureChallenge/newsletters/Water_Nov2006/page5.asp*
*http://www.bchydro.com/guides_tips/green-your-home/water_guide/save_water_in_the_kitchen.html*

Trim your cooking time by using smaller appliances:
*http://www.bchydro.com/guides_tips/green-your-home/appliances_guide/cooking.html*

Research ways to water your lawn and garden more efficiently:
*www.nrcs.usda.gov/feature/highlights/homegarden/lawn.html*

Research the history of lawns and reconsider your choices:
*www.organiclawncare101.com/history.html*

Tips on saving water in your backyard:
*http://www.bchydro.com/guides_tips/green-your-home/water_guide/Save_water_outdoors.html#InGarden*

21. Gershon, David. *Low Carbon Diet* (2007), p.10

| | | PLEDGE DATE | # TIMES / WEEK | # TIMES COMPLETE / 7 WEEKS |
|---|---|---|---|---|

# *INTENTION 10 FURNACE & WATER HEATER EFFICIENCY

Of the roughly 2,500 tonnes of $CO_2$ produced by households on an annual basis, nearly 50% of those emissions are generated from the furnace, and another 20% from the water heater.[22] Implementing the following actions to increase the efficiency of your household's furnace and water heater will reduce emissions, and the heating bill! Doing the actions that refer to adjusting your thermostat alone, can save 1,400 pounds of $CO_2$ per year;[23] the possibilities of reducing your impact are large in this section.

| ACTIONS | The Basics | | | |
|---|---|---|---|---|
| C.10.1 | I will turn down my thermostat during the day and dress more warmly. | | | |
| C.10.2 | I will turn down my thermostat at night while I sleep, and when I leave on a trip. | | | |
| C.10.3 | I will turn the thermostat down by 2 degrees ___ times per week. 🖊 | | | |
| C.10.4 | I will close the dampers and registers in a room where heat is not needed. | | | |

| ACTIONS | I Can Do More | | | |
|---|---|---|---|---|
| C.10.5 | I will turn off the pilot light during the warmer months if my furnace is used solely for home heating. | | | |
| C.10.6 | I will put the water setting to Off or Pilot when I go on a trip. | | | |
| C.10.7 | I will clean my furnace filters this week, so that airflow is not restricted. I will continue to change filters once every 90 days for best air quality. | | | |
| C.10.8 | I will make sure that the warm-air ducts are insulated and the joints are sealed. | | | |
| C.10.9 | I will install a programmable thermostat. | | | |
| C.10.10 | I will insulate my hot water heater with an insulator jacket and exposed pipes with foam sleeves. This will save 175 pounds of $CO_2$ a year.[24] | | | |
| C.10.11 | I will lower the temperature of my hot water heater from high to medium. 🖊 | | | |
| C.10.12 | I will have my furnace tuned up by a professional. If my furnace is more than 15 years old, I will research what it would cost to replace it. *This will reduce my $CO_2$ emissions by 300 pounds a year.*[25] | | | |
| C.10.13 | I will investigate the possibility of installing a tankless (on-demand) water heater. | | | |
| C.10.14 | I will replace my old furnace with an energy-efficient model. This will save 2,400 pounds of $CO_2$ a year.[26] | | | |
| C.10.15 | I will look into the possibility of installing air-to-air heat exchangers or solar hot water system. | | | |

22. Gershon, David. *Low Carbon Diet* (2007), p.21 and p.26
23. Ibid p.26
24. Gershon, David. *Low Carbon Diet* (2007), p. 27
25. Gershon, David. *Low Carbon Diet* (2007), p. 26
26. Ibid p. 27

## INTENTION 10 **FURNACE & WATER HEATER EFFICIENCY**

| | | PLEDGE DATE | # TIMES / WEEK | # TIMES COMPLETE / 7 WEEKS |
|---|---|---|---|---|
| C.10.16 | I will look into haw to draft proof my home. 🌍 | | | |
| C.10.17 | I commit to doing action #(s) ____ consistently for ____ days/weeks/months. | | | |

| Actions | **Educating and Influencing Others** | | | |
|---|---|---|---|---|
| C.10.18 | I will talk to a friend about the environmental and economic benefits of my more energy efficient home. | | | |
| C.10.19 | I will ask a friend, coworker or family to join me in completing action #(s)_____ from this intention. | | | |

| Actions | **Community Projects** | | | |
|---|---|---|---|---|
| C.10.20 | We can personally implement action #(s) _____ from this intention at our school or workplace. | | | |
| C.10.21 | We will investigate into our work-places energy efficiency and personally do what we can to improve it. | | | |
| C.10.22 | We will request for our workplace or school to consider implementing action #(s) _____ from this intention; we will research on how this is done to make it as easy as possible for our company, organization or institute. | | | |

## EXPLORATIONS 🌍

www.bchydro.com/guides_tips/green-your-home/heating_guide.html
www.lowes.com/lowes/lkn?action=howTo&p=Improve/HomeEnergyEfficient.html
Draft proofing your home:
www.bchydro.com/guides_tips/green-your-home/heating_guide/draft_proof_your_home.html

## *INTENTION 11  COOL DOWN MORE EFFICIENTLY

While air conditioning helps us cool off during hot weather, the CO2 emissions from an inefficient use of air conditioning is having the opposite effect on the climate. On average, a room air conditioner generates 1.3 tonnes of $CO_2$ as well as creating a large demand for electricity (and, by extension, fossil fuels).[27] Inefficient air conditioning units lead to greater nitrogen dioxide emissions (smog) and higher energy bills. The following strategies will help keep both myself and the planet cooler during hot weather.

| Actions | **The Basics** | | | |
|---|---|---|---|---|
| C.11.1 | I will close the drapes on windows facing the sun on hot days. | | | |
| C.11.2 | I will turn off my air conditioner once my home is cool. | | | |

27. www.cleanairfoundation.org/keepcool/facts_kc.asp

# INTENTION 11  COOL DOWN MORE EFFICIENTLY

| | | PLEDGE DATE | # TIMES/ WEEK | # TIMES COMPLETE / 7 WEEKS |
|---|---|---|---|---|
| C.11.3 | I will move into cooler rooms of the house during hot weather. | | | |
| C.11.4 | I will keep spaces that don't need cooling, such as closets, closed off when I have the AC on. | | | |
| C.11.5 | I will use my oven less frequently on hot days. | | | |
| C.11.6 | I will inspect the AC unit filter regularly and keep it clean. *If you replace or clean your AC regularly, you save 350 pounds of CO2 a year.* [28] | | | |

| Actions | **I Can Do More** | | | |
|---|---|---|---|---|
| C.11.7 | I will service central AC systems regularly. | | | |
| C.11.8 | I will turn up the temperature setting on my AC to reduce energy use (short sleeve temperature). *If I turn up my thermostat by 4 degrees, I reduce my CO2 emissions by 20 pounds a month.* [29] | | | |
| C.11.9 | I will install a timer on room air conditioners and programmable thermostats on central systems. | | | |
| C.11.10 | I will close the flue of my fireplace when not in use. | | | |
| C.11.11 | I will make sure that my AC unit is the correct size for the amount of space I am cooling and that it is shaded if it has to be positioned in the sun. | | | |
| C.11.12 | I will install a window or ceiling fan. | | | |
| C.11.13 | I will replace my old AC unit with a new Energy Star model. [30] | | | |

| Actions | **Educating and Influencing Others** | | | |
|---|---|---|---|---|
| C.11.14 | I will ask whomever I am getting a ride from, if they don't mind having the window open rather than using the AC in the car. | | | |
| C.11.15 | I will find the person responsible, or I will directly check if our water HVAC system at work is operating at its optimal level. 🌎 | | | |

| Actions | **Community Projects** | | | |
|---|---|---|---|---|
| C.11.16 | We can personally implement action #(s) _____ from this intention at our school or workplace. | | | |
| C.11.17 | We will ask a friend, coworker or family to join us in completing action #(s) _____ from this intention. | | | |

28. Gershon, David. Low Carbon Diet (2007), p. 15
29. Ibid p.15
30. Gershon, David. *Low Carbon Diet* (2007), p. 15

## INTENTION 11  COOL DOWN MORE EFFICIENTLY

| | | PLEDGE DATE | # TIMES/ WEEK | # TIMES COMPLETE / 7 WEEKS |
|---|---|---|---|---|
| C.11.18 | We will request for our workplace or school to consider implementing action #(s) _____ from this intention; we will research on how this is done to make it as easy as possible for our company, organization or institute. | | | |
| C.11.19 | We will get a small group together to implement some of these actions in community, school or workplace. | | | |

## EXPLORATIONS

*www.bchydro.com/guides_tips/green-your-home/cooling_guide.html*
*www.energystar.gov*

Optimal HVAC system operation- HandyCanadian:
*http://www.handycanadian.com/articles-hvac.asp*

## INTENTION 12  SWITCH TO RENEWABLE ENERGY

Using fossil fuels causes extensive damage to the Earth's atmosphere. Choosing to switch to renewable energy sources sends a clear, dual message to government about the importance of developing clean, alternative sources of energy, and the need to greatly reduce our reliance on fossil fuels.[31]

Despite the promise of alternative energy sources -- more appropriately called renewable energy, collectively they provide only about seven percent (7%) of the world's energy needs. This means that fossil fuels, along with nuclear energy -- a controversial, non-renewable energy source -- are supplying 93% of the world's energy resources.

Burning fossil fuels creates carbon dioxide, the number one greenhouse gas contributing to global warming. Combustion of these fossil fuels is considered to be the largest contributing factor to the release of greenhouse gases into the atmosphere. In the 20th century, the average temperature of Earth rose one degree Fahrenheit (1°F).

*http://ecology.com/features/fossilvsrenewable/fossilvsrenewable.html*

| Actions | The Basics | | | |
|---|---|---|---|---|
| C.12.1 | I will contact an energy consultant to explore options currently available to me. (In BC, consider www.citygreen.ca) | | | |

| Actions | I Can Do More | | | |
|---|---|---|---|---|
| C.12.2 | I will research what switching to solar hot water heating or installing photovoltaic solar panels would entail. | | | |
| C.12.3 | I will switch to receiving some of my energy or heating to coming from renewable sources. | | | |

31. Gershon, David. *Low Carbon Diet* (2007), p.29

# INTENTION 12 SWITCH TO RENEWABLE ENERGY

| | | PLEDGE DATE | # TIMES/ WEEK | # TIMES COMPLETE / 7 WEEKS |
|---|---|---|---|---|
| ACTIONS | **Educating and Influencing Others** | | | |
| C.12.4 | I will ask my energy producers if I can switch to renewable energy sources, and if not, I will look into where else I can buy energy. | | | |

| | | | | |
|---|---|---|---|---|
| ACTIONS | **Community Projects.** | | | |
| C.12.5 | We will research into renewable energy companies or organizations and we will volunteer ___ hours for one, or donate ___ dollars towards their cause/projects. | | | |
| C.12.6 | We will implement or advocate for a green renewable energy program for a building we occupy (work, home or community), such as wind or solar energy. | | | |
| C.12.7 | We will get a group together to imagine and brainstorm, what our community might look like in 2030 if it were using 90% less fossil fuels (and having more fun). We will think about how this "came about" and what we can do to make this happen. We will record the process. See Explorations. | | | |

## EXPLORATIONS

Often, it is possible to purchase "green power" from your local utility. This usually costs a little bit extra each month, perhaps $3-$5. However, the money you will be saving on your utilities from conserving energy should balance out. Credit your household 200 pounds of $CO_2$ for each 100 killowat hour of green power you purchase.[32]

Learn more about renewables at *www.davidsuzuki.org/Climate_Change/Energy/Renewables.*

For more information about renewable energy programs in Canada, visit the Government of Canada Office of Energy Efficiency: *www.oee.rncan.gc.ca*

For information on options available in BC, contact The Vancouver Renewable Energy Cooperative (VREC): *www.recov.org*

For articles on hot water and solar technologies, refer to the BC Sustainable Energy Association website: *www.bcsea.org/learn/get-the-facts*

Envisioning a different future:

*Energy Network www.villagevancouver.ca/group/energy1*

Vancouver Peak Oil: *http://vancouverpeakoil.org*

Totnes in Action: *Totnes District 2030, An Energy Descent Action Plan. 2010.*

Zero Waste Vancouver: *http://www.villagevancouver.ca/group/vvwastegroup*

and *http://zerowastevancouver.blogspot.com*

*"You must give birth to your images. They are the future waiting to be born... the future must enter into you long before it happens."* — Rainer Rilke

32. Gershon, David. *Low Carbon Diet* (2007), p. 29

# INTENTION 13 OFFSET MY CARBON FOOTPRINT

Even when I've made lots of good changes, I'm still going to have a carbon footprint. I can help neutralize this by paying into a carbon offset program to finance other activities that reduce or neutralize carbon emissions.[33] It's important to carefully research the various programs that are available. For example, a tree planting program is not effective if the trees are cut down after a few years.

| | | PLEDGE DATE | # TIMES/ WEEK | # TIMES COMPLETE / 7 WEEKS |
|---|---|---|---|---|
| ACTIONS | **The Basics** | | | |
| C.13.1 | I will research and choose a carbon offset program that offers innovative actions that appeal to me. | | | |
| ACTIONS | **I Can Do More** | | | |
| C.13.2 | I will plant ___ trees this week. | | | |
| C.13.3 | I will purchase $ ___ in CO2 emission offsets when I travel by air. If the airline or travel agent doesn't currently offer its customers the option of offsetting their flights, I will ask them to consider it. | | | |
| C.13.4 | I will offset the CO2 produced yearly by my household. | | | |
| C.13.5 | I will offset the CO2 produced yearly by my vehicle(s). | | | |
| C.13.6 | I commit to doing action #(s) ____ consistently for ____ months/years. | | | |
| ACTIONS | **Educating and Influencing Others** | | | |
| C.13.7 | I will encourage a carbon-neutral objective at the next event in which I am involved. 🌐 | | | |
| C.13.8 | I will host the next event I organize as a "carbon neutral" event to bring attention to environmental issues. | | | |
| C.13.9 | I will ask a friend, coworker or family to join me in completing action #(s) _____ from this intention. | | | |
| C.13.10 | I will request for my workplace or school to consider implementing action #(s) _____ from this intention; I will research on how this is done to make it as easy as possible for my company, organization or institute. | | | |
| ACTIONS | **Community Projects** | | | |
| C.13.11 | We will get together with a small group of people, family, friends, coworkers, neighbors or other circle members and we will track and offset the majority of our energy consumption this week. | | | |

33. According to *Low Carbon Diet*, David Gershon, a single tree absorbs one ton (907 kgs) over its life time.

# INTENTION 13  OFFSET MY CARBON FOOTPRINT

| | | PLEDGE DATE | # TIMES / WEEK | # TIMES COMPLETE / 7 WEEKS |
|---|---|---|---|---|
| C.13.12 | We will coordinate a carbon off-setting event, in which we will invite a neighborhood or community to take off-setting action, such as planting trees or fundraising to support a carbon-offsetting organization we believe in. | | | |
| C.13.13 | We will spearhead a carbon neutral institution or workplace initiative.  | | | |

## EXPLORATIONS

Measure the total Carbon Footprint of your home, car, flight, or event and how you can offset it: *http://www.carbonfund.org/Calculators/*

Learn about carbon offsets: *www.davidsuzuki.org/Climate_Change/What_You_Can_Do/carbon_offsets.asp*

Research options available to you at: *www.davidsuzuki.org/Climate_Change/What_You_Can_Do/carbon_neutral.asp*

Learn how to properly plant a tree: *www.treecanada.ca/publications/guide.htm*

David Suzuki's Carbon Neutral events and organizations: *http://www.davidsuzuki.org/what-you-can-do/reduce-your-carbon-footprint/how-to-host-a-sustainable-carbon-neutral-conference-or-other-event/index.php*

*"At first people refuse to believe that a strange new thing can be done. Then they begin to hope it can be done. Then they see it can be done. Then it is done and all the world wonders why it was not done centuries ago."* – Francis Eliza Hodgson Burnett

# RECONNECT TO SELF, OTHERS & EARTH

*I value feeling connected to myself and to others,*
*in human and natural community.*

Many of us live such busy lives that we don't take the time to really get to know ourselves, our neighbors or the web of life we're part of. The Industrial Revolution ushered in a mechanistic world view that assumed we are separate — separate from each other, from other life forms, and from the ecological systems that sustain us all. Yet in the last century, scientific studies of astronomy, quantum physics, systems theory and morphogenic fields are bringing us a new understanding of an evolving, responsive Universe and Earth comprised of dynamic interdependent relationships and systems, not just separate functioning parts.

When we value these relationships we live a richer life. Those who meditate are found to slow down the biological effects of aging, have boosted immunity, and more organized thinking, focus and attentiveness.[11] An interesting article on the effects of living in neighborhoods with trees draws a connection between living greenery and reducing the crime rate— helping residents to be healthier, happier and better able to handle stressful situations[12] Trees and natural landscapes even counter attention deficit disorder.

Being connected within community fosters a feeling of belonging and creates a contagious happiness.[13] Feeling a connection with spirit and the universe deepens meaning in our day-to-day lives. We enrich our life through greater connection with our self, others, our communities, other species, our planet, the universe and spirit. Enjoy this deep and playful value!

## I Intend To:

| | |
|---|---|
| D 1 | **Connect with Yourself** |
| D 2 | **Connect with Others** |
| D 3 | **Connect with Family** |
| D 4 | **Connect with Neighbours** |
| D 5 | **Connect with the Bioregion** |
| D 6 | **Connect with Other Species** |
| D 7 | **Connect with Earth** |
| D 8 | **Connect with the Universe** |
| D 9 | **Connect with the Spirit** |
| D 10 | **Supplement** |

11. http://ezinearticles.com/?Positive-Effects-of-Meditation&id=4851053
12. www.telegraph.co.uk/earth/earthnews/4612176/AAAS-Living-near-trees-makes-people-live-longer-and-feel-happier.html
13. www.actionforhappiness.org/10-keys-to-happier-living/local-community?gclid=CKaY8si8kKoCFQYbQgod1luGyQ

# INTENTION 1 CONNECT WITH YOURSELF

Making time to reconnect with ourselves is a necessity—not a luxury. This time for ourselves helps us to stay in touch with who really are at the core of our being, our inner or essential self, [1] whose dreams, feelings, and wisdom are often brushed aside in order to "fit in" with social circumstances or expectations. The danger of not regularly connecting with ourselves is that we lose touch with our authenticity and, in turn, lose our ability both to create a meaningful life for ourselves and to authentically connect with others in our lives.

| | | PLEDGE DATE | # TIMES/ WEEK | # TIMES COMPLETE / 7 WEEKS |
|---|---|---|---|---|
| ACTIONS | **The Basics** | | | |
| D.1.1 | I will start a journal and will write something in it ___ days this week. | | | |
| D.1.2 | I will set aside ___ minutes for daily meditation, or join a meditation class this week. | | | |
| D.1.3 | I will take ___ hours to explore a creative project ___ day(s) such as writing, painting, music or dance. | | | |
| D.1.4 | I will write a list of all the things I am grateful for this week. | | | |
| D.1.5 | I will write a 'have-do-be', in which I will outline what I want to have in my life, what I want to do, and who I want to be. I will visualize what that looks and feels like. | | | |
| ACTIONS | **I Can Do More** | | | |
| D.1.6 | I will write out a list of simple pleasures. *See Value D Supplement* | | | |
| D.1.7 | I will cut down on the number of hours I work by ___ hours this week. | | | |
| D.1.8 | I will make a list of all the things I loved doing as a child. I will do one of those things. | | | |
| D.1.9 | I will write my own epitaph, (how I would like to be remembered on my gravestone). I will put this up where I will see it regularly, and consider what I can change to fulfill my personal vision. | | | |
| D.1.10 | I will join a group that explores myths, analyzes dreams, utilizes symbols, or other techniques for exploring the subconscious. | | | |
| D.1.11 | I will take time to recognize and develop my own individual gifts, by writing a list of all the things I am good at, all the things I really enjoy doing, and all the things I wasn't good at or didn't enjoy doing, but I have mastered in the course of my lifetime. | | | |
| D.1.12 | I will envision a positive future and read good news publications such as www.yesmagazine.org | | | |
| D.1.13 | I will join a What's Your Tree ~ Be The Change Circle and explore my purpose and passion. www.whatsyourtree.org | | | |

1. The "essential self" is a term coined by Martha Beck in her book *Finding Your Own North Star: Claiming The Life You Were Meant to Live (2001)*.

# INTENTION 1 **CONNECT WITH YOURSELF**

## EXPLORATIONS

Read: Affluenza:

*The All-Consuming Epidemic by John de Graaf, David Wann, and Thomas H. Naylor (Berrett-Koehler Publishers, 2002).*

Be authentic and live "the life that resonates with your inner being".

Read: Being Authentic by L. Mascaro at *http://www.selfgrowth.com/articles/Mascaro1.html*

*"The difference between a six-hour day and an eight-hour day is the difference between energetic, happy people with time to develop their personalities, interests, and relationships and exhausted, harried people who are always racing, always tense, never satisfied, never done."*

# INTENTION 2 **CONNECT WITH OTHERS**

Humans are a social species, by nature. Despite the popularity and endurance of the "rugged individualism" ideology, we cannot go through life without relating to other individuals. It is critically important to both our own and others' well being to support each other and allow ourselves to be supported by others, to show our appreciation for and connect with the people who come into and enrich our lives. Enjoy the opportunity to enrich your own and others' lives by making time for the actions in this section.

| Actions | The Basics | PLEDGE DATE | # TIMES/ WEEK | # TIMES COMPLETE / 7 WEEKS |
|---------|-----------|---|---|---|
| D.2.1 | I will practice being present when spending time with another person. I will not use my phone, turn on the TV or think about what I am doing next, when with another. | | | |
| D.2.2 | I will take the time to get to know an acquaintance better – someone that I normally don't have time for or our transactions are normally more business-like. | | | |
| D.2.3 | I will spend time with someone from a different generation. | | | |
| D.2.4 | I will do something for nothing. | | | |
| D.2.5 | I will smile at strangers __ days this week. | | | |
| D.2.6 | I will choose one person in my community that I perceive as "other", have assumptions about, or would normally feel uncomfortable talking to; I will make an effort to meet that person and get to know them. | | | |
| D.2.7 | I will do the "listen for understanding" practice__ times this week. *See Value D Supplement* | | | |
| D.2.8 | I will have ___ more meals with others. | | | |
| D.2.9 | I will let at least one car in front of me on every journey. | | | |
| D.2.10 | I will take a walk and talk with others. | | | |
| D.2.11 | I will write a letter or email of thanks and appreciation. | | | |

## INTENTION 2  CONNECT WITH OTHERS

| Actions | I Can Do More | PLEDGE DATE | # TIMES / WEEK | # TIMES COMPLETE / 7 WEEKS |
|---|---|---|---|---|
| D.2.12 | I will spend time with an elder and ask to hear their life stories, recognizing and appreciating their contributions to my life. | | | |
| D.2.13 | I will spend time with a child. I will let go of any other agenda than being in the moment and experience life through their eyes. | | | |
| D.2.14 | I will practice compassion meditation every day for one week. *See Value D Supplement* | | | |
| D.2.15 | I will let all my actions flow with a consciousness of others. I will allow all of my interactions with others to come from a place of compassion, including relating to others I feel I have nothing in common with. | | | |
| D.2.16 | I will bake something for a friend. | | | |
| D.2.17 | I will offer up my seat on the bus once this week. | | | |
| D.2.18 | I will learn one good joke and tell it a few times. | | | |
| D.2.19 | I will volunteer my time doing some kind of direct service (e.g., serving meals in a soup kitchen) and undertake this service with an attitude of love and selflessness. | | | |
| D.2.20 | I will learn to say something friendly in another language. | | | |
| D.2.21 | I will write my will and use it to good effect. | | | |
| D.2.22 | I will register online as an organ donor. | | | |
| D.2.23 | I will hug _ person(s) _day(s) this week, a different person everyday! | | | |
| D.2.24 | I will "pass it forward" once this week: a book, sweater, CD gifting to a friend or stranger. | | | |
| D.2.25 | I will research local youth groups in my community looking for volunteers or resources and commit to helping out. | | | |
| D.2.26 | I will research and explore the philosophy and values of a faith different than my own. | | | |
| D.2.27 | I will visit a faith centre, place of worship, or spiritual teaching that is different from my upbringing. | | | |
| D.2.28 | I will visit with someone of a different faith background. | | | |
| D.2.29 | I will register for a first-aid course this week. | | | |
| D.2.30 | I will recycle my glasses to Seva or an organization in my area that will repair and redistribute them to those in need. | | | |

## INTENTION 2 **CONNECT WITH OTHERS**

| | | PLEDGE DATE | # TIMES / WEEK | # TIMES COMPLETE / 7 WEEKS |
|---|---|---|---|---|
| D.2.31 | I will learn more about a different culture this week: _____, and share my findings with the Circle. | | | |
| D.2.32 | I will give blood. | | | |
| D.2.33 | I will give my change to charity ___ times. | | | |
| D.2.34 | I will pledge to give what is asked of me, without judgment. | | | |
| D.2.35 | I commit to doing action #(s) ____ consistently for ____ days/weeks/months. | | | |
| Actions | **Educating or Influencing Others** | | | |
| D.2.36 | I will ask a friend, coworker or family to join me in completing action #(s)_____ from this intention. | | | |
| Actions | **Community Projects** | | | |
| D.2.37 | We will start a youth group or other community group. | | | |
| D.2.38 | We will bring to light unfair treatment of individuals or groups. We will listen to and record their stories and find some way of sharing them. | | | |
| D.2.39 | We will host or organize a workshop or speaker in which we are interested. | | | |
| D.2.40 | We will research and explore the philosophy and values of different faiths and visit difference faith centres to explore the common values and unique spiritual expressions of humanity. | | | |
| D.2.41 | We will choose some of the Actions listed above and do them as a group, sharing our varying experiences with each other. | | | |

## EXPLORATIONS 🌍

*www.liveyourlifewell.org*

*www.spiritualityandpractice.com/practices/practices.php?id=4*

Building community through food:

- Host a traditional Sunday dinner
- Potluck with friends
- Start your own gourmet club
- Plan a cooking party
- Consider joining a cooking cooperative

# INTENTION 3 CONNECT WITH FAMILY

Kinship within our immediate and extended families goes beyond our genetic ties; family members play a role in shaping our life stories and our temperaments, and our lives are often enriched through a special connection with one or more of our relatives. In many cultures, one can usually count on immediate and extended family for social and practical support, but in many of the westernized nations, the emphasis on independence and the insular "nuclear family"[2] unit has weakened these ties. Choose to reverse this trend by reaching out and connecting to both immediate and extended family members of various generations and take pleasure in renewed kinship bonds.

| Actions | The Basics | | | |
|---|---|---|---|---|
| D.3.1 | I will spend time with each member of my family and build strong relationships, and make the effort to become a genuine part of each other's lives. | | | |
| D.3.2 | I will appreciate my family life and enjoy the people I love. | | | |
| D.3.3 | I will spend time with a relative from another generation this week. | | | |
| D.3.4 | I will make family dinners sacred. | | | |

| Actions | I Can Do More | | | |
|---|---|---|---|---|
| D.3.5 | I will write letters of gratitude to different members of my family, regardless of the current status of our relationship. I will recall every kind thing that person has done for me, everything I have learned from them, and how they helped to shape me, positively, into who I am today. I will consider sending the letters, reading them to a friend or loved one, or just keeping them for myself. | | | |
| D.3.6 | I will take my kids or those in my family on hikes, to the beach, and other outdoor activities as an antidote to the electronic world. | | | |
| D.3.7 | Many cultures have recorded the details of their history in stories that they tell each other and pass down from generation to generation. This week, I will sit down with my family and create a story that tells our family history, as we know it. | | | |
| D.3.8 | I will choose one senior family member and ask them to tell me their most vivid life memories. I will record these conversations and create a compilation of their life stories. | | | |
| D.3.9 | I will hold a family reunion or gathering. | | | |

## EXPLORATIONS

Research what naturalists' clubs are available in your community. For BC, visit Wild BC at *www.hctf.ca/wild/about.htm and Young Naturalist Club at www.ync.ca*
Bike Programs: *http://www.bikesense.bc.ca/courses.htm*
Youth engagement: *www.greenlearning.ca/*

2. *A nuclear family* is a family group consisting of a father and mother and their children, who share living quarters *(en.wikipedia.org/wiki/Nuclear_family)*, whereas an extended family includes grandparents, aunts, uncles, cousins.

# INTENTION 4 CONNECT WITH NEIGHBOURS

How well do you know your neighbours? Are you friends with them and comfortable 'borrowing an egg' when you need to, like in days gone by? Or do you nod "hello," but have never introduced yourself to them or learned their names? Connecting with neighbours offers many opportunities to enrich lives through shared experiences—whether social activities or working together to bring about improvements to the neighbourhood—or through finding ways to support each other in meeting practical needs from helping with gardening or babysitting to carpooling or sharing meals.

| Actions | The Basics | PLEDGE DATE | # TIMES/ WEEK | # TIMES COMPLETE / 7 WEEKS |
|---|---|---|---|---|
| D.4.1 | I will walk around my neighbourhood more and chat "over the fence". | | | |
| D.4.2 | I will canvass for a worthy cause in my neighbourhood and introduce myself as I go along. | | | |
| D.4.3 | I will share more with my neighbours. Babysitting, tools, meals, carpooling, dog walking, assisting seniors, fresh produce, etc. | | | |
| D.4.4 | I will attend ___ community events hosted by my local schools, churches, and community centre this week. | | | |

| Actions | I Can Do More | | | |
|---|---|---|---|---|
| D.4.5 | This week, I will offer to mow an elder neighbour's lawn. | | | |
| D.4.6 | I will grow something and get to know my fellow gardeners. | | | |
| D.4.7 | I will learn about the history of my community this week. | | | |
| D.4.8 | This week I will interview someone who lived in my neighbourhood at least 40 years ago. How have things changed? | | | |
| D.4.9 | I will contact my local residents association this week for ideas on how I can help nurture community where I live. | | | |
| D.4.10 | I will research the most pressing social issues of my larger community: wealth and poverty; ethnic, cultural and religious discrimination; employment and unemployment, etc. | | | |
| D.4.11 | I will identify the indigenous ancestors of my area. What did they eat? What were their shelters like? What stories did they tell? | | | |
| D.4.12 | I will research the current living conditions of the indigenous people of my area. | | | |
| D.4.13 | I commit to doing action #(s) ___ consistently for ___ days/weeks/months. | | | |

## INTENTION 4  CONNECT WITH NEIGHBOURS

| | | PLEDGE DATE | # TIMES/ WEEK | # TIMES COMPLETE / 7 WEEKS |
|---|---|---|---|---|
| ACTIONS | **Educating and Influencing Others** | | | |
| D.4.14 | I will become a member of my residents' association this week and get involved in civic affairs. | | | |
| D.4.15 | I will find a creative way to share the knowledge I learned doing action #_____, with my neighbours or greater community. | | | |
| D.4.16 | I will help to organize community events: artists' walks, garden tours, cultural festivals, neighbourhood fairs, Holiday craft sales, May Fairs, etc. | | | |
| D.4.17 | At work, I will send a greeting note to the other local businesses saying who we are, and organize a time for them to come over and say hi. | | | |
| D.4.18 | I will ask a friend, coworker or family to join me in completing action #(s)_____ from this intention. | | | |
| ACTIONS | **Community Projects** | | | |
| D.4.19 | We will organize a project with our neighbours, such as gather unused household hazardous wastes to deliver to the proper disposal site, start a community garden, car free day, etc. | | | |
| D.4.20 | We will host a social gathering this month that brings neighbours together: a block party, garage sale, BTC Information meeting, etc. | | | |
| D.4.21 | We will create a neighbourhood directory or initiate a Blockwatch program this week. | | | |
| D.4.22 | We will take on action # ___ to connect with our workplace neighbours. | | | |
| D.4.23 | We will organize a work event and invite the other local businesses to create connections and networking with our work neighbours. | | | |

## EXPLORATIONS

Find out if a resident's association exists in your area and if so, review their website and newsletters for information on what is going on in your neighbourhood. Look for tour information on urban agriculture sites in your area. (For Vancouver, BC, go to *www.cityfarmer.org*) Journal on how you might protect the vigour and diversity of your local community, not only human connections but all the non-human natural world as well – the local flows of air, water and animals which move through the land in your home place.

*"Once the truth has made its presence felt in a single soul, nothing can ever stop it from invading everything and setting fire to everything."* – Teilhard de Chardin

# INTENTION 5  CONNECT WITH THE BIOREGION

How much do you know or notice about the bioregion in which you live? Taking the time to get to know as much as possible about the bioregion in which you live will do more than just expand your general knowledge about where you live—it will also forge a more personal connection with and concern for your bioregion, helping you to become an empowered environmental steward of your "corner of the world". Make a choice to look into and follow through on some of these actions to initiate a connection with your bioregion.

| Actions | The Basics | | | |
|---|---|---|---|---|
| D.5.1 | I will walk to the store instead of driving. What do I see, feel, hear, and smell that I didn't notice before? | | | |
| D.5.2 | I will learn the names of all the natural geographic features of my local bioregion: rivers, mountains, bays, inlets, etc. | | | |
| D.5.3 | I will map my place, imagining that someone was coming to visit me from a distant place. I'll provide written directions to my house without referring to any streets, buildings, or other human artifacts. | | | |
| D.5.4 | I will visit the local botanical gardens and explore native plant information. | | | |
| D.5.5 | I will choose a spot in my neighbourhood that will be my special spot where I can sit and observe nature and people for one hour. I will record my observations. | | | |
| D.5.6 | I will take a walk led by naturalists this week to learn more about local species. | | | |
| D.5.7 | I will start a daily observation journal this week and document a natural event every day. Example: The leaves have turned crimson on the maple tree. | | | |

| Actions | I Can Do More | | | |
|---|---|---|---|---|
| D.5.8 | I will learn as much as I can about my bioregion – a unique place with its own watershed, soils, climate, plants, animals, and history. | | | |
| D.5.9 | I will research into what resources could sustain me with my bioregion, or what can be cultivated here from energy, to building products, beautiful places, local manufactures, minerals and agricultural potential. | | | |
| D.5.10 | I will make a meal this week entirely from foods grown in my bioregion. | | | |
| D.5.11 | If I grew up in a different bioregion, I will write a list of observations about the differences between where I grew up and where I now live. Example: The temperature, the seasons, the amount of rain or sun, the type of earth, the birds, the plants, the insects. | | | |

Column headers: PLEDGE DATE | # TIMES/ WEEK | # TIMES COMPLETE / 7 WEEKS

# INTENTION 5  CONNECT WITH THE BIOREGION

| Actions | **Educating and Influencing Others** | PLEDGE DATE | # TIMES/ WEEK | # TIMES COMPLETE / 7 WEEKS |
|---|---|---|---|---|
| D.5.12 | I will prepare a meal for my friends or family with food that has been grown in my bioregion. | | | |
| D.5.13 | I will share with friends or my Circle research I completed on features of our bioregion. | | | |
| D.5.14 | I will do some research and present to my Circle the possibility of having a guest speaker for an extra session to help us better discover our bioregion. | | | |
| D.5.15 | I will ask a friend, coworker or family to join me in completing action #(s)_____ from this intention. | | | |

| Actions | **Community Projects** | | | |
|---|---|---|---|---|
| D.5.16 | We will join with other community members to help maintain common spaces (parks or sidewalks), or help clean up a local beach. | | | |
| D.5.17 | We will contact the Parks Department and local chapters of conservation groups to inquire about on-going restoration activities in our community. We will volunteer to help with activities: Clearing non-native vegetation, rehabilitating streams for aquatic life, protecting wetlands, collecting seeds, etc. | | | |
| D.5.18 | We will plan and host a bio-region social with presentations and opportunities for elders to share stories that give us a deeper understanding of the region. | | | |

## EXPLORATIONS

Find organizations in your locality that educate people about their bioregion. Your local Audubon Society chapter or parks and recreation department may be a good source of information. In Vancouver, BC, visit:
*http://www.naturevancouver.ca*
Research what community cleanup activities are held annually in your area. For ideas in Vancouver, BC, try
*http://www.city.vancouver.bc.ca/residents.htm or www.evergreen.ca*

Organizations that participate in cleaning up garbage from shorelines:
- Canada BC Pacific Streamkeepers Foundation: *www.pskf.ca/*
- Canada Nation-wide: The Great Shoreline Cleanup: *www.vanaqua.org/cleanup/*
- USA Streamkeeper Foundation: *www.streamkeeper.org/*

*"The earth is not a collection of objects, but a communion of subjects."* – Thomas Berry

# INTENTION 6  CONNECT WITH OTHER SPECIES

We are often so busy with our lives that we do not slow down enough to really notice or learn about the plants and animals in our own area. By taking the time to observe and study the animals and plants in our area, we can form a connection with these other life forms and start to acquire a deeper sense of an interdependent relationship between ourselves and our surroundings.

| Actions | The Basics | Pledge Date | # Times/ Week | # Times Complete / 7 Weeks |
|---|---|---|---|---|
| D.6.1 | This week, I will get to know a wild flower, bird, or mammal native to my area, and learn to understand what role it plays in Earth's living systems. | | | |
| D.6.2 | I will learn to identify and locate an edible plant that grows wild in my area. | | | |
| D.6.3 | I will provide food and water for birds and take time to observe their behaviour. | | | |
| D.6.4 | I will be with an old tree on a regular basis and get to know it. I will sit under it; ask it questions; build a relationship. | | | |

| Actions | I Can Do More | | | |
|---|---|---|---|---|
| D.6.5 | I will dig into one square foot of soil in a garden or compost pile, and count how many living creatures I observe. | | | |
| D.6.6 | If I have a cat, I will put 3 bells on my cat's collar to prevent it from hunting migratory birds. | | | |
| D.6.7 | I will learn to identify and locate a medicinal plant that grows wild in my area. | | | |
| D.6.8 | I will learn about the "weeds" that grow in my area. Which are edible? Which are medicinal? | | | |
| D.6.9 | I will learn about the endangered species in my region. Acquire the Endangered Species Toolkit from Sierra Club, or another local source. 🌐 | | | |
| D.6.10 | I will thank the plants and animals that I eat, before each meal this week. | | | |
| D.6.11 | I will choose an animal species to 'adopt' and support an organization that is working on its behalf. | | | |
| D.6.12 | I will learn about the extinction crises of all large wild animals (lion, tiger, elephant, and silver back gorilla). | | | |
| D.6.13 | I will financially support an organization trying to preserve wild animals and their habitat. | | | |
| D.6.14 | I will join the SPCA or other organization assisting abandoned domestic pets this week. | | | |
| D.6.15 | I commit to doing action #(s) ____ consistently for ____ days/weeks/months. | | | |

## INTENTION 6  CONNECT WITH OTHER SPECIES

| | | PLEDGE DATE | # TIMES/ WEEK | # TIMES COMPLETE / 7 WEEKS |
|---|---|---|---|---|
| ACTIONS | **Educating and Influencing Others** | | | |
| D.6.16 | I will ask a friend, coworker or family to join me in completing action #(s)_____ from this intention. | | | |

## EXPLORATIONS

Find what kind of field guides that have been written for your area: birds, plants, sea life, stars, weather and so on.

Try to find one at a local second-hand bookstore or library.

Learn about the sentient responses of plants and "Morphogenic Fields" as researched by Dr. Rupert Sheldrake, Ph.D:

*www.sheldrake.org/homepage.html*

Learn about the effects of domestic cats on the wild bird populations at:

*www.abcbirds.org/abcprograms/policy/cats/materials/predation.pdf*

Review the Sierra Club Endangered Species Toolkit at:

*http://www.sierraclub.bc.ca/endangered-species*

## INTENTION 7  CONNECT WITH EARTH

We often get so busy and focused on attending to the daily chores and requirements of the mundane aspects of our lives that we do not stop to consciously recognize or acknowledge our deep connection with the Earth, forgetting that we are made from the same elements and chemicals. We may also forget that we are completely dependent on Earth's resources to keep us alive, and in turn that it is our responsibility to care for Earth, as well as the animals and plants that share the planet with us. Incorporating some of the following actions into our daily or weekly practices will help forge or re-establish our connection with Earth, and express our gratitude for all the gifts we receive.

| ACTIONS | **The Basics** | | | |
|---|---|---|---|---|
| D.7.1 | I will start a new practice that reminds me of my connection to the natural surroundings. Example: I might trace the source of my food before starting the evening meal, or go outside each morning to hear which birds are singing or to see what the clouds are doing. | | | |
| D.7.2 | I will find new ways to connect with nature, such as participating in nature walks in my community or rediscovery camps in regions outside of my city or town. | | | |
| D.7.3 | I will establish a new ritual of giving thanks: to the sun for giving light and warmth, to the plants for providing food, to the trees for providing paper and wood, to the ancient decayed plants for providing fuel. | | | |
| D.7.4 | I will practice grounding myself; I can do this by sitting or standing barefooted on the grass and visualize myself putting roots deep into the earth and my branches and consciousness reaching high into the sky or heavens. | | | |

# INTENTION 7 **CONNECT WITH EARTH**

| | | PLEDGE DATE | # TIMES/ WEEK | # TIMES COMPLETE / 7 WEEKS |
|---|---|---|---|---|
| ACTIONS | **I Can Do More** | | | |
| D.7.5 | I will research and explore First Nations beliefs about connecting with nature. | | | |
| D.7.6 | I will take a Contemplative Nature Walk. *See Value D Supplement* | | | |
| D.7.7 | I will experience Being Seen by Nature. *See Value D Supplement* | | | |
| D.7.8 | I will do Partnering with Nature: A Discovery Experience. *See Value D Supplement* | | | |
| D.7.9 | I will do Experience with Nature: Being There. *See Value D Supplement* | | | |
| D.7.10 | I will consider Brian Swimme's perspective of gravity (the allurement of planetary beings and that Earth is holding us to her with love). To experience this, I will try the following exercise: lie on your back on a starry night and imagine yourself looking down at the stars, with the love of mother Earth holding you close. | | | |
| D.7.11 | I will plant something and nurture its growth and well-being, understanding the importance of gardening as a healing experience and tool. | | | |
| D.7.12 | I will plan a vacation that will immerse me in a unique aspect of nature, such as kayaking in Haida Gwaii, or rafting remote rivers of North America, or hiking in mountains, or experiencing a natural wonder like the Grand Canyon. | | | |
| D.7.13 | I commit to doing action #(s) ____ consistently for ____ days/weeks/months. | | | |
| D.7.14 | I will join the SPCA or other organization assisting abandoned domestic pets this week. | | | |
| D.7.15 | I commit to doing action #(s) ____ consistently for ____ days/weeks/months. | | | |

| | | | | |
|---|---|---|---|---|
| ACTIONS | **Educating and Influencing Others** | | | |
| D.7.16 | I will invite a friend or family member to join me in an activity that I find grounding and connective to our planet, such as hiking, Tai Chi, skinny dipping, etc. | | | |
| D.7.17 | I will ask a friend, coworker or family to join me in completing action #(s)_____ from this intention. | | | |

| | | | | |
|---|---|---|---|---|
| ACTIONS | **Community Projects** | | | |
| D.7.18 | I will volunteer with an environmental organization in my area and participate in a wilderness or nature restoration project. | | | |
| D.7.19 | I will organize gatherings on solstices or equinox, and make sure to honour and give thanks for the many earthly pleasures and gifts. | | | |

## INTENTION 7 CONNECT WITH EARTH

### EXPLORATIONS 🌍

Read: Michael Cohen's Reconnecting with Nature: Finding Wellness Through Restoring Your Bond with the Earth (Ecopress, 2007) and continue to deepen your relationship with Earth.

Learn about the Gaia Hypothesis, and the theory of Earth as a self-regulating being by NASA scientist Dr. James Lovelock. *http://www.ecolo.org/lovelock*

Watch a short video on the documentary Call to Life: Facing the Mass Extinction at: *http://www.speciesalliance.org/video.php*

## INTENTION 8 CONNECT WITH THE UNIVERSE

Western civilizations once believed that the Earth was the centre of the universe and that other planets and stars revolved around the earth. Just a few decades ago we thought the Universe was static and only recently have scientists recognized that the Universe is an expanding and responsive energy field.[3] To consider the multitude of miracles that collectively create the evolution of our planet, in our solar system, in our spiral galaxy the Milky Way, comprised of hundreds of millions of stars larger than our sun, in a Universe of hundreds of millions of galaxies, that apparently flared forth 13.7 billion years ago as an ever expanding paradigm of time and space... this can be a truly AWE-some experience, evoking a sense of both Mystery and humility. Deep gratitude may prevail. You might choose to reconnect with the Universe by exploring some of the following actions.

| Actions | The Basics | | | |
|---|---|---|---|---|
| D.8.1 | I will remember our place in the galaxy by going outside at dusk and imagine the earth is rolling backwards away from the sun, or at dawn and imagine the earth is rolling forward toward the sun. | | | |
| D.8.2 | Watch the video clips www.globalmindshift.com under Expand your Mind: Who Am I? Birth to Earth: Who Am I? Life to Human: The New Story. | | | |
| D.8.3 | I will watch Cosmic Zoom to appreciate the grandeur of our "home" galaxy at http://www.youtube.com/watch?v=VgfwCrKe_Fk | | | |

| Actions | I Can Do More | | | |
|---|---|---|---|---|
| D.8.4 | I will lie on my back, gazing up at the stars at night, then I will switch my perspective to recall that I am being held onto ground by Earth's gravity, and I will look DOWN at the stars and feel grateful for the love that holds me close to my home planet. | | | |
| D.8.5 | I will go to a planetarium and learn more about the galaxies and stars. | | | |
| D.8.6 | I will read The Universe Story by Thomas Berry and Brian Swimme. | | | |

3. Berry, Thomas and Brian, Swimme. *The Universe Story (1992).*

## INTENTION 8  CONNECT WITH THE UNIVERSE

| | | PLEDGE DATE | # TIMES/ WEEK | # TIMES COMPLETE / 7 WEEKS |
|---|---|---|---|---|
| D.8.7 | I will watch the Powers of the Universe DVD series of lectures by Brian Swimme. | | | |
| D.8.8 | I will read The Universe is a Green Dragon, by Brian Swimme. | | | |
| ACTIONS | **Educating and Influencing Others** | | | |
| D.8.9 | I will initiate a contemplative discussion about the universe, our role in it and its potential expansiveness. | | | |
| D.8.10 | I will ask a friend, coworker or family to join me in completing action #(s)_____ from this intention. | | | |

## EXPLORATIONS

Read: Michael Cohen's Reconnecting with Nature: Finding Wellness Through Restoring Your Bond with the Earth (Ecopress, 2007) and continue to deepen your relationship with Earth.

Learn about the Gaia Hypothesis, and the theory of Earth as a self-regulating being by NASA scientist Dr. James Lovelock. *http://www.ecolo.org/lovelock*

Watch a short video on the documentary Call to Life: Facing the Mass Extinction at: *http://www.speciesalliance.org/video.php*

# INTENTION 9  CONNECT WITH SPIRIT

In today' society we are encouraged to constantly move forward, doing, achieving, never content. Most of our actions are concerned either with preparing for the future or dwelling on the past. Rarely do we slow down enough to simply appreciate the present moment. As a result, it is often difficult to feel any sort of connection with others, with nature or even with ourselves.

When we learn to find stillness in our bodies and in our thoughts, we discover a deep knowing of our interconnectedness and wholeness, just as we are. By abiding in the present moment, we reconnect with our True Self and to Universal Consciousness itself. We soon realize that we are not physical beings with occasional spiritual experiences, but a spiritual being in a physical form in order to serve our purpose here.

Once we experience this inner, contented Self, we can draw upon this strength, even when the external world is full of chaos. The more we remain present and conscious in everything we do, the more we perceive divinity in everything around us.

The path to remembering our spiritual connection is different for each of us, and it is an ongoing journey. You may wish to choose one of the actions listed below and to delve deeply into it for as long as you need to.

# INTENTION 9  CONNECT WITH SPIRIT

| Actions | The Basics | PLEDGE DATE | # TIMES/ WEEK | # TIMES COMPLETE / 7 WEEKS |
|---|---|---|---|---|
| D.9.1 | I will practice BEING in the PRESENT. | | | |
| D.9.2 | I will practice Stillness and Silence for ___ minutes every day this week. *See Value D Supplement* | | | |
| D.9.3 | I will practice Non-Judgment for __ hours __ days this week. *See Value D Supplement* | | | |
| D.9.4 | I will detach from my viewpoint being 'reality' ___ times this week. | | | |
| D.9.5 | I will practice Self-Nurturance and Nurturance of Others __ times this week. *See Value D Supplement* | | | |
| D.9.6 | I will consider my Life's Purpose for ___ hours this week. | | | |

| Actions | I Can Do More | | | |
|---|---|---|---|---|
| D.9.7 | I will practice Yoga __ days a week for __ weeks. | | | |
| D.9.8 | I will choose one day this week (____ ) to give a small gift to every person I encounter. *See Value D Supplement* | | | |
| D.9.9 | I will practice Non-Judgment & Non-Labeling for a whole week. *See Value D Supplement* | | | |
| D.9.10 | I will practice Gratitude. *See Value D Supplement* | | | |
| D.9.11 | I will practice Non-Reaction and Witness Consciousness.[4] | | | |
| D.9.12 | I will practice Non-Blaming.[5] | | | |
| D.9.13 | I will let go of my need to Control.[6] | | | |
| D.9.14 | I will be open to moments of Divinity. | | | |
| D.9.15 | I commit to doing action #(s) ____ consistently for ____ days/weeks/months. | | | |

| Actions | Educating and Influencing Others | | | |
|---|---|---|---|---|
| D.9.16 | I will start a conversation with someone about what it means to them to be present and connected to Spirit. I will be open to understanding and sensitive to not imposing my views. | | | |
| D.9.17 | I will ask a friend, coworker or family to join me in completing action #(s)_____ from this intention. | | | |

4. Chopra, Deepak. The Seven Spiritual Laws of Success: A practical Guide to the Fulfillment of Your Dreams. San Rafael, CA: New World Library
5. Ibid
6. Ibid

## INTENTION 9 **CONNECT WITH SPIRIT**

### EXPLORATIONS 🌐

Chopra, Deepak. *The Seven Spiritual Laws of Success: A practical Guide to the Fulfillment of Your Dreams.*
San Rafael, CA: New World Library

Tolle, Eckhart. *A New Earth: Awakening to Your Life's Purpose.* New York, NY: Penguin Group, 2005.

Dyer, Wayne. *Being in Balance.* London, England: Hay House, 2006.

Dyer, Wayne. *The Power of Intention.* London, England: Hay House UK, 2004.

Cope, Stephen. *Yoga and the Quest for the True. Self.* New York, NY: Bantam, 2000.

## VALUE D SUPPLEMENT

## INTENTION 1 **CONNECT WITH SELF – SIMPLE PLEASURES**

Your own sense of happiness is something you can directly contribute to, in simple ways that develop your self-love. In your journal, make a list of all the things you like that bring you joy in the following sensory categories. Keep some space at the end of each category for adding to later as they come to mind.

- Smell
- Taste
- See
- Hear
- Touch/Feel

Choose one thing from each category to give to yourself this week. Consider choosing one thing from each category to give to yourself next week.

Write down a few things that were important to you about this activity.

Write down what good feelings may have been brought on by doing this activity.

How would you feel if your ability to experience them was taken from you?

Do you trust these feelings as being real? What effect does this activity have on your sense of self worth?

## INTENTION 2 **CONNECT WITH OTHERS – LISTEN FOR UNDERSTANDING**

When we aren't pre-occupied with whether we agree with another person when they are communicating, we can be much more open to hearing correctly what they are trying to communicate. There are five easy steps to this practice that are worth learning. The communication is between a Sender and a Listener and both can use this practice to understand each other clearly.

### Be Self-Aware.

Our 1st step is to be self-aware of the Context, Information, Understanding and Personal Screen that I have as sender and you have as listener. My CONTEXT is my state of mind and frame of reference. For example, I got up early, in a great mood with this beautiful sunny day; or I'm late and stressed...

### The Information is the Raw Data.

People have different kinds and amounts of pertinent information.Each person's UNDERSTANDING of this information will vary depending upon how long they've been using it or teaching it to others. It is easy to assume that another's understanding is the same as my own, so I have to be careful not to leave out important clarifications.

We must also be aware that our communicating is going through our PERSONAL SCREEN. This is the total of all my experiences in life and the meaning I give it. Everyone's screen is different. With my personal screen I can unconsciously delete data, distort data and inappropriately generalise. The tighter my personal screen, the harder it is for me to understand another's point of view. The larger the space in my screen and the more curious I am of other, the more information I can take in.

### Listen for Understanding

The 2nd step is LISTENING for clarity and understanding. We use "reflective listening" to reflect back to each other what we have heard. This is where we can pick up that we might have left out some important information. We learn to selfregulate how much information we give at once, checking to ensure that the other has received it correctly. We use "I" statements taking personal responsibility for our assumptions and understanding.

The 3rd Step is where we come to an UNDERSTANDING of what's being communicated. In this process we recognize different points of view we may have, and any issues or differences are made clear. We check out any assumptions we may have made, and we can also see where we have agreement.

### Action

Based on this understanding we come to Step 4 and we take the appropriate ACTIONS. We agree on a process to work through any differences from a place of understanding each other's needs.

### Revisions

In Step 5 we assess our actions and make the appropriate REVISIONS. We use this entire communication process again for Step 5, when we communicate clearly and listen for understanding what's working well, what needs improvement, and what's just right.

## INTENTION 2 CONNECT WITH OTHERS – COMPASSION MEDITATION

Compassion, or Metta Meditation, involves silently repeating certain phrases that express the intention to move from judgment to caring, from isolation to connection, from indifference or dislike to understanding. You don't have to force a particular feeling or get rid of unpleasant or undesirable reaction; the power of the practice is in the wholehearted gathering of attention and energy, and concentrating on each phrase.

The traditional Buddhist phrases are: "May you be free of pain and sorrow. May you be well and happy." This does not mean that we wish the world will not visit us with adversity, for we have no control over that. But in this phrase we recognize that we do not have to be attached to the experience such that we cause ourselves pain and suffering.

You can begin with a 20-minute session and increase the time gradually until you are meditating for half an hour at a time. If your mind wanders, don't be concerned. Notice whatever has captured your attention, let go of the thought or feeling, and simply return to the phrases. If you have to do that over and over again, it is fine.

To begin, take a comfortable position. You may want to sit in a chair or on cushions on the floor (just make sure your back is erect without being strained or overarched). You can also lie down. Take a few deep, soft breaths to let your body settle.

Closing your eyes or leaving them slightly open, start by thinking of someone you care about already – perhaps they have been good or inspiring to you. You can visualize this person or say their name to yourself, get a feeling for their presence, and silently offer phrases of compassion and loving kindness: "May you be free of pain and sorrow. May you be well and happy."

After a few minutes, shift your attention inward and offer the phrases of compassion to someone you find difficult. Get a feeling for the person's presence, and offer the phrases of compassion to them.

Then, after some time, turn to someone you've barely met – someone at work you don't know, the checkout grocer or delivery person. Even without knowing their name, you can get a sense of the person, perhaps an image, and offer the phrases of loving kindness to them.

And now, turn your attention inward and offer the phrases of compassion to yourself. "May I be free of pain and sorrow. May I be well and happy." Sadly westerners often find this the most difficult part of the meditation.

We close with the offering of compassion to people everywhere, to all non-human creatures, all sentient beings and all forms of life, without limit, without exception: "May all beings be free of pain and sorrow. May all be well and happy."

## INTENTION 7 **CONNECT WITH EARTH – CONTEMPLATIVE NATURE WALK**

Take a moment to mark the beginning of your walk at the trailhead with a prayer, moment of silence or other simple ceremony. Then begin. Keep your life question (eg: What is my purpose?) in your heart and on your lips as often as you can. Your task is to simply walk, notice and see what nature offers as symbols, signs and possible metaphors. This requires a whole different way of looking and seeing and listening. Stay open to everything and consider the possibility that signs and messages are everywhere. Ask the trees, ask the birds, ask the river and listen deeply. Have a small pad and pen. Stop occasionally to jot down key messages for journaling. When your journey is complete, stop at the trailhead ending again and offer a moment of gratitude for whatever you have experienced.

If you are with a group, reconvene to share your experiences. If you are alone, take this time to journal your experience, so that you can reflect upon it and share it with the group at another time.

Mark the beginning and end of the walk with a simple ceremony.

Nature is both a teacher and healer if we open ourselves to its wisdom and voices. Keep your life question, such as "What is my purpose?" or "what message do you have to share with me?" in your heart and on your lips as often as you can. Ask it of specific natural features such as a tree, rock, bird, and also ask of nature in general...and listen deeply.

Your task is to simply walk, notice and see what nature offers as symbols, signs and possible metaphors. This requires a whole different way of looking and seeing and listening. You are entering an allegorical landscape – a microcosmic life story that is a mirror of your own psyche. Stay open to everything and the symbols of your inner journey will be reflected to you. Treat the entire walk as if it were a dream in which everything has meaning and everything represents the self.

There is a relationship between mindscape and landscape, between inner and outer. When you truly look, you dissolve the boundaries. Be aware of nature observing you. Move back and forth between being a human observing nature (self) and then nature observing yourself.

Get all your senses completely involved: taste, smell, hear, see, and touch. Through the senses, we experience what it truly means to be alive. Perception is a participatory act – remember the relationships you engage in when you simply look. Pay attention to energy – both your own and the subtle energies of the natural world. Let those energies draw you in. By seeing differently, we do differently.

*"We can now think of vision as a deeply reciprocal event, a participatory activity in which both the seer and the seen as dynamic players. To see is to interact with the visible, to act and be acted upon. It is participation in the ongoing evolution of the visible world." David Abram*

## INTENTION 7  CONNECT WITH EARTH – BEING SEEN BY NATURE

Over the next week, find time to have this experience in nature and write about it in your journal. Go to your 'special place' or any other site in nature. Simply 'be' in nature and relax into the place. Check in with yourself and take note of the state of your being. Breathe.

After a few minutes, shift your awareness to being a "human being observing nature". Use all your human senses to simply observe nature fully. What do you see? What do you smell? What do you hear?

After a few minutes shift your attention to being "observed by nature, observed by the "soul of the world". During this step see if you can feel the eyes and ears of nature upon you. Are the trees watching you? Can the wildlife nearby sense your presence? Is the ground aware of you? Let yourself be fully immersed in both the idea and the felt experience of being seen. Is there a mutual sense of seeing going on?

Come back to some kind of middle ground where you are both observing and being observed by the natural world. Can you feel the relationship between your soul and that of the world? Can you relate to the idea of kinship with the world of nature? Please write about this experience in your journal.

## INTENTION 7  CONNECT WITH EARTH – PARTNERING WITH NATURE – A DISCOVERY EXPERIENCE

Go to somewhere or something in nature that you like, that you find attractive. A park, a backyard, an aquarium, or a potted plant will do. When you get to it, notice how you feel. Can you thank it for your good feelings? Now, treat this area fairly, with respect, as an equal or friend. Don't bully it, instead gain its consent for you to visit and enjoy it. Ask this natural area for its permission for you to be there. Doing this increases your sensitivity to the area. Ask it if it will help you learn from it. You cannot learn if you are going to injure or destroy it, or it you. Wait for about half a minute. Look for adverse signals of danger such as thorns, bees, cliff faces, etc. If the area still feels attractive, or becomes more attractive, you have gained its consent. If this portion of the natural area you visit no longer feels attractive, simply select another natural part that attracts you and repeat this process. Do this until you find an area where a safe attraction feeling remains for 10 seconds, then thank it.

Once you have gained an area's consent, compare how you feel about being there now with how you felt about it when you first arrived. Has any change occurred? If you find that gaining a natural area's consent to visit it is rewarding, remember that you can do this whenever you want to become more connected to nature. Here are some reactions of past participants to this activity. Add your reaction(s) to the list. Share them with others: "It was hot. Soon after I asked for permission to be with the grove of young trees, a gentle, refreshing breeze came through them. It cooled me, and the trees waved their leaves at me. It felt good, like the grove smiled its consent." "I was attracted to the sound of a raven on the rocks ahead. I stopped and sought its consent for me to enjoy its presence. It began to come closer and closer, increasing my fun and excitement. That was unforgettable." Write down important things you learned from this activity. Write down what good feelings may have been brought on by doing this activity. To discover this activity's effect on your sense of self, complete a sentence that begins with: "My experience in Nature shows me that I am a person who gets good feelings from _____." Use this sentence form whenever you connect with a natural attraction. It reinforces a positive, natural self-image.

## INTENTION 7 **EXPERIENCE WITH NATURE – BEING THERE**

The idea that nature and our inner nature contain intelligent love frightens many of us. In chapter five, we will learn that we can know and learn through five natural senses. Each is a natural intelligence. The nature-separated way that we learn to think overrides them.

To know how to swim, sooner or later you must want to get into the pool. To truly enjoy nature's intelligence, you must want to experience your natural senses. Cognitive awareness of them alone is not enough. Why not get in the pool now? Feel them. Go to a nearby attractive natural area, the more natural the better. If necessary, a potted plant, an aquarium, or a square foot of lawn or earth will do. Ask for the natural area's permission for you to visit and become involved with it. Gain its consent to help you with this activity (reference Partnering with Nature for this process).

The natural world involves a type of love that neither uses nor understands verbal language. Can part of you trust that? That part is your inner nature. To help discover and enjoy non-verbal nature including your inner nature, take ten minutes or more to enjoy this natural area's attractions in silence. Each attraction is a mini-love.

During this ten minutes ask the natural area: "Who are you without your names and labels?" Wait for some kind of response to come into consciousness. Then ask: "Who am I without my name?" Again wait for some response to occur. Repeat this procedure for ten minutes or more as long as it is attractive. Write down what you experienced during this visit and its value. Do you know this natural place and/or yourself differently now than before you did this activity? Recognize that you may have converted the experience into words, but the natural area never said a word.

## INTENTION 9 **RE-CONNECT WITH SPIRIT**

### Stillness and Silence

Start with just a few minutes a day, and work up to 10-20 minutes of silence each day. This may take the form of simply sitting and being, or it may evolve into a meditation practice. The idea is to become comfortable with just BEING, with non-doing.

### Non-Judgement

For this day, focus on your mind's reaction to each event that occurs. Be aware of thought-forms that arise. For instance, what is your internal dialogue when you first wake in the morning, and while you eat a meal, and during your commute, and in your conversations with others? Notice if there are any patterns to how your mind reacts to events that occur during your day. Notice how often your mind judges or labels your experiences as good or bad, comfortable or uncomfortable, right or wrong. Journal about your experience.

### Non-Judgement and Non-Labelling

Try to accept people and situations simply as they are in that moment, without having to put a label on it. Given our deep conditioning, at first this work will be to constantly remind ourselves not to judge. A statement or mantra such as "Today I will not judge" or "Practice acceptance" may be helpful. Listening for understanding instead of agreement is also helpful in this practice.

### Self Nurturance

By filling our own cup as often as needed, doing things that nourish our heart and soul, we will be more able to give freely to others. Let go of any guilt conditioning and replace it with benevolent sharing of joy and contentment with others.

### Gift to Everyone

A smile, a compliment, a silent blessing, verbal appreciation, affection... consider what you would like to receive, and, most importantly, consider what you think this person would like to receive.

### Yoga

Start by 'dropping-in' and taking a class. Try different styles of classes until you find one that suits you. Explore the more subtle aspects of yoga such as connection with your breath, concentration, meditation and relaxation practices, in addition to the physical practice.

### Practice Gratitude

Open to receive gifts that are offered us, by others and by the world. Once we practice giving with joy, it is easier to understand the gifts offered to us with joy. If uncomfortable receiving praise or compliments or affection, notice the mind's resistance and then try to open as fully as possible. If you feel guilt for having too much, remember that "giving and receiving are different aspects of the flow of energy in the universe". Start a daily gratitude practice, either in a journal or by sharing with a loved one at the beginning or ending of each day, or as a meditation.

### Affirmation Prayer

A simple prayer or affirmation repeated throughout the day can help set the mind's programming in a positive direction. Create one that works for you. Eg. *"Divine Wisdom empowers me to make wise choices."*

# SUPPORT A GREEN & JUST SOCIETY

*I value standing up for what I believe in, and supporting a socially just human presence on this planet.*

It saddens us when we see a friend or family member being disrespected, abused or hurt by factors out of their control. We want to reach out to support or stand up to this injustice. Abuse and injustice are happening all over the world, yet when it isn't right in front of us, we don't always feel its urgency. Or we may feel too much urgency of too many injustices, leaving us overwhelmed, and apt to turn away.

The economic inequalities are present and daunting: over three billion people live on less than $2/day. That's almost half the world's population. At least 80% of our population lives on less than $10 a day.[14] Vulnerable communities are being hit harder and sooner by the environmental degradation caused by human activity and climate change.[15]

Violence, oppression and disempowerment of women remains prevalent in our world; estimates say that one in every three women has been beaten, coerced into sex or otherwise abused in her life.[16] We import a good majority of our fruits, vegetables and nuts,[17] yet our local farmers are shutting down because of development and consumer choices.

When we recognize the interconnection of all life, we can see how each action and decision affects a multitude of other species and people. Many of our daily choices can either help or further degrade millions of people and our over-stressed environment. There are ways to be informed and make educated purchases, sharing our knowledge with others— and we can reach out to our greater community by influencing those in power and giving what resources (time, financial support, items, skills) we can provide.

## I Intend To:

E 1     **Become an Agent of Change**

E 2     **Support Local Farmers and Public Water**

E 3     **Engage the Community, Government & Industry**

E 4     **Get Informed & Stay Informed**

E 5     **Align Shopping Practices with Values**

E 6     **Be a Socially & Environmentally Responsible Investor**

E 7     **Promote Gender Equality**

E 8     **Promote Economic Equality**

E 9     **Supplement**

14. www.globalissues.org/article/26/poverty-facts-and-stats
15. www.feps-europe.eu/fileadmin/downloads/globalprogressive/1005_ClimateChangeSocialJustice_LL.pdf
16. www.wecanbc.ca
17. www.phsa.ca/NR/rdonlyres/C3E70150-66FF-48E1-B2F1-45BB83B03019/0/FoodforThought_ResearchBooklet_FINAL.pdf

# INTENTION 1 BECOME AN AGENT OF CHANGE

Large scale changes often occur when a "critical mass" has been reached. This means enough people have decided to adopt sustainable living habits to make a noticeable difference, and have joined together to influence the political will of elected politicians. How do we create a critical mass? By talking to others about environmental and social justice issues, telling them about the sustainability practices we have embraced, the benefits we are seeing, and encouraging them to try out some of these practices for themselves.

| | | PLEDGE DATE | # TIMES/ WEEK | # TIMES COMPLETE / 7 WEEKS |
|---|---|---|---|---|
| ACTIONS | **The Basics** | | | |
| E.1.1 | I will develop the habit of awareness to ask myself: "Does this choice support a socially just and sustainable society?" | | | |
| E.1.2 | I will check out various organizations to learn what others in my community are doing, with the intention of being inspired by the major efforts being made at all levels of society. | | | |
| E.1.3 | I will donate ____ dollars to the environmental and social justice organization(s) of my choice. | | | |
| ACTIONS | **I Can Do More** | | | |
| E.1.4 | I will make plans to start a Be The Change Action Circle. | | | |
| E.1.5 | I will connect with other Be The Change Circles and explore common interest for collective action. | | | |
| E.1.6 | I will join my local Transition Town group. See www.villagevancouver.ca or other transition town groups. | | | |
| E.1.7 | I will host an Awaken the Dreamer Change the Dream Symposium at work or at a social organization to help activate more people to join the movement. | | | |
| E.1.8 | I will train to be a presenter of the Awaken the Dreamer Symposium. | | | |
| E.1.9 | I will become a member of GreenPeace and support their efforts financially $___ and/or with ___ hours of volunteer time. | | | |
| E.1.10 | I will join a local organization that inspires me, and get involved in their activities. | | | |
| ACTIONS | **Educating and Influencing Others** | | | |
| E.1.11 | I will make green and just living the norm by talking to ____ people about what I'm learning in this program and encouraging them to lower their ecological impact. | | | |
| E.1.12 | I will send ___ e-mail messages with green tips for sustainable living to my friends. | | | |

# INTENTION 1  BECOME AN AGENT OF CHANGE

| | | PLEDGE DATE | # TIMES/ WEEK | # TIMES COMPLETE / 7 WEEKS |
|---|---|---|---|---|
| E.1.13 | I will identify all the people I know who might be receptive to lowering their ecological impact, and invite them to an information evening about an Action Circle. | | | |
| E.1.14 | I will ask a friend, coworker or family to join me in completing action #(s) _____ from this intention. | | | |

| Actions | **Community Projects,** | | | |
|---|---|---|---|---|
| E.1.15 | We will commit ___ volunteer hours to an environmental or social justice organization of our choosing. | | | |
| E.1.16 | We will lobby local and provincial governments to invest in programs that create green jobs. | | | |
| E.1.17 | We will volunteer to take on the responsibilities of a Sustainability Coordinator at our workplaces. | | | |
| E.1.18 | We will implement a program in order to recognize, reward, and publicize our good performance and innovation in sustainable practices at our workplaces or at organizations of which we are a part. | | | |
| E.1.19 | We will spearhead committees to explore sustainable manufacturing and eco-innovation for the products that our companies sell. | | | |
| E.1.20 | We will brainstorm with our Action Circle the issue that we want to work on together and set our goals for the next 8 sessions. | | | |

## EXPLORATIONS

Explore resources at *www.bethechangecircles.org*

Research what is being done to address global warming in your country.

Visit One Earth Initiative: *www.oneearthweb.org*

The Canadian Living Planet Report 2008: *assets.panda.org/downloads/living_planet_report_2008.pdf*

Read: *Blessed Unrest: How the Largest Movement in t he World came into Being and Why No One Saw It Coming*

*(The Penguin Group, 2007) by Paul Hawken* for an account of the accelerated movement of civil society world-round.

Join the global movement underway to find solutions.

Sustainability Coordinator Information: *See the "The Green Workplace" by Leigh Stringer.*

*The Green Workplace.com: www.thegreenworkplace.com/*

Sustainable Manufacturing and Eco-innovation – *The Observer, OECD Policy Brief*

*"Whatever you do, or dream you can, begin it. Boldness has genius, power and magic in it. Begin it now."* – Johann Wolfgang Von Goethe

## INTENTION 2 SUPPORT LOCAL FARMERS

Do you notice where your food comes from when you buy it? Have you ever wondered about our vulnerability within the international food system? The topic of 'Food Security' has become of great interest and taken more seriously by more and more people. One way to increase our food security and reduce the use of fossil fuels is to eat more locally. Choosing to buy locally produced food is a great way to increase demand and production of local goods. It is a smart choice for the environment, the local economy, your health and your taste buds. Produce that is locally grown is much better tasting and often more nutrient-dense because it is harvested at its ripest. Small scale local farms often use sustainable growing practices that inflict less damage on the environment, as well as producing less $CO_2$ in transport. Along with supporting local food, the common right to public drinkable water is also becoming a contentious issue. We can support public water over the privatization of water by our daily actions and by supporting campaigns that stand up for others that are greatly affected by this struggle.

| Actions | The Basics | PLEDGE DATE | # TIMES/ WEEK | # TIMES COMPLETE / 7 WEEKS |
|---------|-----------|---|---|---|
| E.2.1 | I will find out about local farmers' markets in my community this week. | | | |
| E.2.2 | I will buy local produce in season rather than produce imported from other countries. | | | |
| E.2.3 | I will find alternative places to purchase local, sustainable food. 🌐 | | | |
| E.2.4 | I will carry a reusable water bottle and use tap or filtered water rather than purchase bottled water. | | | |
| E.2.5 | I will find out where my foods come from when I shop. | | | |

| Actions | I Can Do More | | | |
|---------|---------------|---|---|---|
| E.2.6 | I will buy fresh foods from farmers' markets or directly from a farmer ___ times this week. | | | |
| E.2.7 | I will support local food cooperatives, by becoming a member or endorsing one. | | | |
| E.2.8 | I will visit a farm near my city at harvest time and pick my own fruits and vegetables. | | | |
| E.2.9 | I will research into various bottled water companies and see where their water comes from. | | | |
| E.2.10 | I will research into the effects of the privatization of bottled water and the human rights issues behind it. 🌐 | | | |
| E.2.11 | I will join Food and Water Watch's "Take Back the Tap" campaign at: www.TakeBackTheTap.org | | | |
| E.2.12 | I commit to doing action #(s) ___ consistently for ___ days/weeks/months. | | | |

# INTENTION 2  SUPPORT LOCAL FARMERS

| | | PLEDGE DATE | # TIMES/ WEEK | # TIMES COMPLETE / 7 WEEKS |
|---|---|---|---|---|
| ACTIONS | **Educating and Influencing Others** | | | |
| E.2.13 | I will lobby grocers to provide local and sustainable food. If your grocer isn't available to speak with, drop-off a pre-prepared request card on your way out. There are pre-written request cards for grocers available at www.sustainabletable.org. | | | |
| E.2.14 | I will invite friends, family or neighbours over for a local meal and discuss this issue with them. | | | |
| E.2.15 | I will suggest a bio-regional organic potluck for our next organizational meeting, party or event rather than ordering out or catering. | | | |
| E.2.16 | I will ask a friend, coworker or family to join me in completing action #(s) _____ from this intention. | | | |

| | | | | |
|---|---|---|---|---|
| ACTIONS | **Community Projects,** | | | |
| E.2.17 | We will ensure that our workplaces do not use plastic water bottles. 🌐 | | | |
| E.2.18 | We will research sustainable and healthy options for our water source and present our findings to a decision-maker at our workplaces. | | | |
| E.2.19 | We will volunteer to work on an organic farm this year. | | | |
| E.2.20 | We will look into developing initiatives to bring local food into our schools and/ or workplaces this week. 🌐 | | | |
| E.2.21 | We will lobby our schools and/or workplaces to be bottled water free. | | | |
| E.2.22 | We will join a local community agricultural program this week. csafarms.ca/index.html | | | |

## EXPLORATIONS 🌐

Find out about local sustainable food. In BC this information can be found at *www.EatWellGuide.org*.

To find your local farmer's market in BC go to *www.bcfarmersmarket.org* or in Vancouver BC go to *www.eatlocal.org/*

More information:

Find farmer markets in your neighbourhood, including local churches that have organic produce delivered on Sundays.

For Vancouver, BC, visit: *www.getlocalbc.org/en/* or *www.vancouverfarmersmarket.com*

Start a local buying-club or join an Organic Produce Cooperative: *www.nowbc.ca/*

Learn about urban organic home delivery companies in your city: *www.spud.ca*

World Food Security: Food and Agriculture Organizations (FAO) of the United Nations: *www.fao.org*

Food Security: *www.foodsecurecanada.org*
*www.100milediet.org/why-eat-local*
*www.sustainabletable.org/intro/3steps/#act*
*www.sustainabletable.org/schools/dining*
*www.bchydro.com/guides_tips/green_your_business/food_health_guide.html*

PLEDGE DATE

# TIMES/ WEEK

# TIMES COMPLETE / 7 WEEKS

## INTENTION 2  SUPPORT LOCAL FARMERS

Consider joining the Slow Food International Movement: *www.slowfoodvancouver.com*

Stolen Harvest: *The hijacking of the global food supply, by Vandana Shiva, South End Press, 2000.*

Privitization of water:

*Blue Gold: The Battle Against Corporate Theft of the World's Water Maude Barlow & Tony Clarke (2002).*

*Whose Water Is It? The Unquenchable Thirst of a Water Hungry World Bernadette McDonald & Douglas Jehl (eds.) (2003).*

BC water updates: *http://www.cupe.bc.ca/campaigns/water-watch/*

Plastic Water Bottle Information- Toxic Free Canada: *http://www.toxicfreecanada.ca/campaign.asp?c=11*

Safe Water Bottles- Toxic Free Canada: *http://www.leas.ca/on-the-trail-of-water-bottle-toxins.htm*

## INTENTION 3  PARTICIPATORY DEMOCRACY

Being an agent of change is easier when you can team up with others who are also committed to implementing sustainable practices in order to preserve our world for future generations. There are lots of constructive methods to work for change, such as: joining organizations that inform the public about the issues, creating public support to push for a change in political priorities, lobbying elected officials at all levels of governments, and writing letters to the editors of newspapers and politicians. This section focuses on the power of small groups of people that can do amazing things together and the importance of influencing those in positions of power and influence.

| ACTIONS | The Basics | | | |
|---|---|---|---|---|
| E.3.1 | I will investigate what sort of regular events are organized in my community to bring people together to discuss social and environmental issues and attend one this week. | | | |
| E.3.2 | I will align myself with others that are passionate about being green and socially just; this will give me the support I need to do the work I am passionate about. 🌍 | | | |
| E.3.3 | I will support the campaign of a social justice or environmental organization and help lobby our government to make this a better place today and for future generations. | | | |
| E.3.4 | I will check out a political party websites this week: record whom I can contact; find out when and where they will be speaking; and show up to an event to ask prepared questions relating to social justice and environmental sustainability. | | | |

| ACTIONS | Educating and Influencing Others | | | |
|---|---|---|---|---|
| E.3.5 | To get more involved, I will find out if there is a sustainability and/or social justice committee in my area of work and see how we can work together to educate or influence my co-workers or community. | | | |
| E.3.6 | I will write a letter to the editor of a local or national newspaper. 🌍 | | | |

# INTENTION 3  **PARTICIPATORY DEMOCRACY**

| | | PLEDGE DATE | # TIMES/ WEEK | # TIMES COMPLETE / 7 WEEKS |
|---|---|---|---|---|
| E.3.7 | I will get to know who the key politicians are and practice participatory democracy. I don't have to be an expert to have a say. I will write a letter to express my opinions on the current issues. | | | |
| E.3.8 | I will ask a friend, coworker or family to join me in completing action #(s) _____ from this intention. | | | |
| E.3.9 | We will follow municipal plans to develop urban infrastructure that will enhance sustainable living. We will go to a public meeting this month. | | | |
| E.3.10 | We will write to an electronics company to inquire about the toxins in their products, and express our concerns about pollution, worker exposure during production, recycling, and disposal. | | | |

## EXPLORATIONS

Networking and meeting like minded people: there are many things happening that connect people that are like minded and want to talk sustainability lifestyles and that support each other. Green Drinks *http://www.greendrinks.org/BC/ Vancouver* is a way to chat over drinks, joining Village Vancouver *www.villagevancouver.ca/* and attending any of their events, joining the Vancouver collective housing yahoo group *vancollectivehousenetwork.blogspot.com/*, and there are many more ways of connecting to others that support a more environmentally conscious and socially just lifestyle.

Voters Taking Action for Climate Change (VTACC) is a citizen's organization in Vancouver, BC. Some of their Political Action Primer and letter writing advice is included at the end of Value D. For more great ideas check out their website: *www.vtacc.org*

Use the internet to compile a list of your political representatives, school trustees, city councilors, provincial and federal representatives. For example: Provincial *(www.gov.bc.ca/contacts/)* and Canada, *(www2.parl.gc.ca/Parlinfo/ Compilations/HouseOfCommons/MemberByPostalCode.aspx?Menu=HOC )*

For tips on getting a politician's attention, visit: *www.davidsuzuki.org/NatureChallenge/What_is_it/The_actions/Get_attention.asp*

See Value E Intention III Supplement on page 122 for a quick primer on "How to get your message in the news media" and other important skills for citizens working in the public arena

## INTENTION 4  GET INFORMED & STAY INFORMED

Being an agent of change is easier when you can team up with others who are also committed to implementing sustainable practices in order to preserve our world for future generations. There are lots of constructive methods to work for change, such as: joining organizations that inform the public about the issues, creating public support to push for a change in political priorities, lobbying elected officials at all levels of governments, and writing letters to the editors of newspapers and politicians. This section focuses on the power of small groups of people that can do amazing things together and the importance of influencing those in positions of power and influence.

# INTENTION 4 **GET INFORMED & STAY INFORMED**

| Actions | The Basics | PLEDGE DATE | # TIMES / WEEK | # TIMES COMPLETE / 7 WEEKS |
|---------|-----------|-------------|----------------|----------------------------|
| E.4.1 | I will choose a social justice issue or topic that I am interested in and will research it, including the different perspectives on it and stakeholders involved. I will also stay current on the issue. | | | |
| E.4.2 | I will support the country's national public radio and television broadcast (CBC in Canada, NPR in the U.S.), and seek out my city's Cooperative radio station(s). | | | |
| E.4.3 | I will read the Earth Charter at www.earthcharter.org and discuss this with my Circle. | | | |
| E.4.4 | I will subscribe to The Canadian Centre for Policy Alternatives at: www.policyalternatives.ca | | | |
| E.4.5 | I will subscribe to the Common Sense Canadian email newsletter: http://thecanadian.org/. | | | |
| E.4.6 | I will check out alternative press, such as The Guardian and Truth Out. | | | |

| Actions | I Can Do More | | | |
|---------|---------------|--|--|--|
| E.4.7 | I will review National Geographic Society's Green Guide at www.thegreenguide.com/ | | | |
| E.4.8 | I will read about David Suzuki's Nature Challenge at: www.davidsuzuki.org/NatureChallenge | | | |
| E.4.9 | I will watch Oxfam International's films on social and eco justice at: www.oxfam.org/en/video | | | |
| E.4.10 | I will watch the documentary "Manufactured Landscapes", following photographer Edward Burtynsky on a tour of China, focusing on industrial subjects. | | | |
| E.4.11 | I will watch movies on www.GlobalOnenessProject.com | | | |
| E.4.12 | I will watch the 20-minute video The Story of Stuff http://www.storyofstuff.com | | | |
| E.4.13 | I will read the GreenPeace Green Living Guide. | | | |
| E.4.14 | I will find out what a "Sweatshop" is and where they are found. | | | |
| E.4.15 | I will ask for recommendations on books that address social justice issues and read one. | | | |
| E.4.16 | I will check out the offerings of publishers such as New Society Publishers at www.newsociety.com/NSPhome.php | | | |
| E.4.17 | I will make arrangements to attend the Bioneers Conference in California, or one of their local broadcasting events. | | | |

# INTENTION 4  GET INFORMED & STAY INFORMED

| | | PLEDGE DATE | # TIMES / WEEK | # TIMES COMPLETE / 7 WEEKS |
|---|---|---|---|---|
| ACTIONS | **Educating and Influencing Others** | | | |
| E.4.18 | I will find ways to implement the Earth Charter, and seek to apply it to my organization or personal life, to the best of my ability, and promote it actively to others. | | | |
| E.4.19 | I will start a discussion with a friend, or acquaintance, asking about their opinion on a social justice issue, and sharing some of what I have recently learned. | | | |
| E.4.20 | I will introduce a book into a book club that has social or environmental significance. | | | |
| E.4.21 | I will start a blog, where I will address environmental or social justice issues. | | | |
| E.4.22 | I will ask a friend, coworker or family to join me in completing action #(s) _____ from this intention. | | | |

| | | | | |
|---|---|---|---|---|
| ACTIONS | **Community Projects** | | | |
| E.4.23 | We will start a Study Action Circle that will discuss and stay current on a predetermined social or ecological matter. | | | |

# INTENTION 5  ETHICAL SHOPPING

As consumers, we can send out clear messages about what we value by our power to vote with our dollars. One way we can do this is to buy products or services only from companies that are clearly committed to incorporating sustainable practices into their businesses. This means spending some time researching different companies to verify whether their business practices do in fact align with your values, or whether the company is merely resorting to "green washing" or slick marketing claims to sell its merchandise.[1]

| | | | | |
|---|---|---|---|---|
| ACTIONS | **The Basics** | | | |
| E.5.1 | I will learn the difference between buying "green", "certified-organic" "fair-trade", "cruelty-free", and "local", and what that means for the environment and social-justice. | | | |
| E.5.2 | I will use "the Better World Shopping Guide" by Ellis Jones to become a socially responsible shopper. | | | |
| E.5.3 | I will check out the Good Guide http://www.goodguide.com/ to discover which products are more or less ethical, healthy and environmentally friendly. | | | |
| E.5.4 | I will read the labels on everything I buy so I know where my purchases come from. | | | |
| E.5.5 | I will shop at locally-owned stores and those known for good environmental practices. | | | |

1. As there are no regulations around the use of labels such as "natural" or "ethical" in advertising, it is all the more important to ask questions and look for recognized certifications. "Green washing" refers to an advertising strategy that creates the illusion that a product is "green" through claims that are essentially meaningless; e.g., a product is free of a certain ingredient when the ingredient is not used anyway, or is legally prohibited from being used.

# INTENTION 5  **ETHICAL SHOPPING**

| | | PLEDGE DATE | # TIMES/ WEEK | # TIMES COMPLETE / 7 WEEKS |
|---|---|---|---|---|
| ACTIONS | **I Can Do More** | | | |
| E.5.6 | I will research into the top 5 brands or stores that I purchase my goods from, and examine their environmental and social ethics. 🌐 | | | |
| E.5.7 | I will research the Lifecycle Impact (from cradle to grave) of ____# of products I purchase, to examine its ecological impact and human health risks. 🌐 | | | |
| E.5.8 | I will only buy things that I am sure are produced, purchased and distributed in a way that aligns with my values for ____ long. | | | |
| E.5.9 | I will assess what percentage of my personal care products are green and cruelty-free, and find alternatives. | | | |
| E.5.10 | I will buy fair trade and reject anything that causes oppression of others. | | | |
| E.5.11 | I commit to doing action #(s) ____ consistently for ____ days/weeks/months. | | | |
| ACTIONS | **Educating and Influencing Others** | | | |
| E.5.12 | I will ask store managers to stock organic, local, and fair trade certified product | | | |
| E.5.13 | I will make a point of thanking local merchants for stocking products in alignment with my values or products or companies that I have requested. | | | |
| E.5.14 | I will give someone a membership to an environmental organization in lieu of a manufactured gift. | | | |
| ACTIONS | **Community Projects** | | | |
| E.5.15 | We will research products used in our workplaces, and, if any goods demonstrate a negative Lifecycle Impact, we will research alternative products with better assessments. We will bring our results to a decision-maker and request positive change. | | | |
| E.5.16 | If we are aware of any non-local purchased goods, we will research a local supplier that might offer comparable products. We will bring our research to a decision-maker at work. | | | |

## EXPLORATIONS

For company rankings on social responsibility: *www.betterworldshopper.org*

For info on 'cradle to grave' impact of products: *http://en.wikipedia.org/wiki/life_cycle_assessment*

*"Everyone thinks of changing the world,*

*but no one thinks of changing himself."* – Leo Tolstoy

# INTENTION 6  BE A RESPONSIBLE INVESTOR

In addition to voting with our dollars when it comes to purchasing clothing, food, or other items, we can consider the social implications of where we bank, and what kinds of companies we invest with (or have money invested on our behalf). Ensuring that our invested money aligns with our values may take some time, but when enough individuals opt for ethical investment funds, it sends out a clear message that we will not support corporations who harm the environment or support war.

| Actions | The Basics | Pledge Date | # Times/ Week | # Times Complete / 7 Weeks |
|---|---|---|---|---|
| E.6.1 | I will invest in companies that I feel are making an effort to be green, and avoid companies involved in activities with which I do not agree. | | | |
| E.6.2 | I will make sure I believe in the projects and companies that my banking institute invests in, and if not, I will switch to another institute that aligns with my values. | | | |
| E.6.3 | I will research the topic of micro financing. | | | |

| Actions | I Can Do More | | | |
|---|---|---|---|---|
| E.6.4 | I will join a cooperative store in my community. | | | |
| E.6.5 | I will put a portion of any investment income into a socially and environmentally responsible mutual fund. | | | |
| E.6.6 | I will dedicate ___ hours this week to learning about Ethical Investment Funds and Socially Responsible Investing. | | | |
| E.6.7 | I will invest into a micro loan. | | | |

| Actions | Educating and Influencing Others | | | |
|---|---|---|---|---|
| E.6.8 | I will ask friends or family that have investments about the companies they support and if/why they believe in supporting them. | | | |
| E.6.9 | I will ask for gifts for various occasions to come in the form of micro financing or investing in socially and environmentally just organizations rather than receiving material items. | | | |
| E.6.10 | If I decide to leave a financial institution or investment broker for a more environmental and socially just choice, I will let them know why and ask them to contact me if things change. | | | |
| E.6.11 | I will ask a friend, coworker or family to join me in completing action #(s) _____ from this intention. | | | |

## INTENTION 6   BE A RESPONSIBLE INVESTOR

| Actions | Community Projects | | | |
|---|---|---|---|---|
| E.6.12 | We will find out if the companies we work for have set up sustainability committees to monitor the company's performance, and if not, we will lobby for one. | | | |
| E.6.13 | We will encourage our workplaces to adopt pension plans that do not invest in companies that harm the environment or in any way support war. | | | |

## EXPLORATIONS

Research "socially responsible investing":

www.amnesty.ca/business/sri.php  and http://www.greenmoneyjournal.com

Explore what is meant by "full-cost accounting" or "true-cost accounting", where costs and advantages over the lifetime of the product (cradle-to-grave) are considered in terms of environmental, economical and social impacts.

Learn more about how we can build a sustainable economy:

http://www.davidsuzuki.org/publications/reports/2001/canada---climate-change-and-the-new-economy/index.php

## INTENTION 7   PROMOTE GENDER EQUALITY

It is important to promote equity between men and women, with regard to benefiting from social and economic infrastructures, services and general welfare. We must reduce and eventually eliminate the differences in opportunity between men and women in receiving health, education, employment, and involvement in economic activities and decision-making mechanisms. Gender inequality is one of many social issues that have become exacerbated by environmental degradation and climate change.

| Actions | The Basics | | | |
|---|---|---|---|---|
| E.7.1 | I will research into one of the organizations mentioned below to better inform myself of gender equality issues and their connection to environmental issues, such as global warming. | | | |
| E.7.2 | I will support fair-trade items made in developing countries, and products made by women with small business loans in poor countries. Example: www.hopeforwomen.com | | | |
| E.7.3 | I will visit or write to ___ local government representatives, letting them know that women's rights and climate change are important issues for Canadian citizens. | | | |
| E.7.4 | I will become a member of Oxfam at www.oxfam.ca and support their Climate Justice Campaign "Stop Harming, Start Helping: Women's Rights and Climate Change. | | | |

# INTENTION 7 PROMOTE GENDER EQUALITY

| | | PLEDGE DATE | # TIMES / WEEK | # TIMES COMPLETE / 7 WEEKS |
|---|---|---|---|---|
| ACTIONS | **I Can Do More** | | | |
| E.7.5 | I will get involved with Sierra Club's Global Population and Environment Program at www.sierraclub.org/population, and join their Population Justice Environmental Challenge Campaign. | | | |
| E.7.6 | I will join the "Think Outside the Bottle" Campaign at www.stopcorporateabuse.org/think-outside-bottle. | | | |
| E.7.7 | I will join the "We Can (End All Violence Against Women)" Campaign (www.wecanbc.ca) and become a "Change Maker." | | | |
| E.7.8 | I will get involved with the Feminist Majority Foundation's (FMF) Campaign for Afghan Women and Girls at www.helpafghanwomen.com. | | | |
| ACTIONS | **Educating and Influencing Others** | | | |
| E.7.9 | I will have a discussion with other people, about the issue of violence against women. | | | |
| E.7.10 | I will promote education about the link between climate change and human rights. I will write ___ letters to news agencies and governments. | | | |
| E.7.11 | I will write a letter to my government asking them to honour their Millennium Development Goal commitments at www.endpoverty2015.org | | | |
| E.7.12 | I will have a discussion with other people about how woman living in poverty, in many developing nations, are bearing the largest burden for climate change RIGHT NOW, yet have little responsibility for the position they are in. | | | |
| E.7.13 | I will ask a friend, coworker or family to join me in completing action #(s) _____ from this intention. | | | |
| ACTIONS | **Community Projects** | | | |
| E.7.14 | This month, we will research and write an article for our local newspaper regarding the issues surrounding violence against women. Alternately, we will publish the article on a blog or online news journal. | | | |
| E.7.15 | We will organize a film night around the issue of gender inequality this month. | | | |
| E.7.16 | We will contact our local women's centre or transition house and organize a fundraiser, or find other ways we can volunteer our time and energy to help out. | | | |
| E.7.17 | We will commemorate the Montreal Massacre on December 6th by distributing white ribbons. See www.whiteribbon.ca/about_us/ | | | |
| E.7.18 | We will get involved with a local initiative to support global change initiatives, specifically: _____ | | | |

# INTENTION 7 **PROMOTE GENDER EQUALITY**

| PLEDGE DATE | # TIMES/ WEEK | # TIMES COMPLETE / 7 WEEKS |
|---|---|---|

## EXPLORATIONS

Research the organizations that are relevant in your community on the internet. For example:

- Status of Women in Canada: *www.swc-cfc.gc.ca*
- BC Ministry of Health, Women's Issues: *www.health.gov.bc.ca/women-and-children/womens-issues/*
- The National Organization for Women (NOW): *www.now.org/index.html*
- Pathways of Women's Empowerment: *www.pathwaysofempowerment.org*

Learn more about women and water scarcity with the Feminist Majority Foundation's toolkit "Breaking the Cycle: Women, Water, and the Search for Equity" *www.feminist.org*

Find out about Women's Environment and Development Organization (WEDO) at *www.wedo.org* and take action using their online toolkit to advocate for gender and climate change

Find out more about the Global Gender and Climate Alliance (GGCA) at: *www.wedo.org/learn/campaigns/climatechange/global-gender-and-climate-alliance*

Research the Women and Climate Change Campaign through www.FeministCampus.org at: *www.feministcampus.org/know/global/womenclimatechange*

Watch the DVD "Killing us Softly", about the advertising industry and how it affects society's views of women and girls: *www.youtube.com/watch?v=QSXDCMSlv_l http://www.youtube.com/watch?v=9zudgbjFvvo&feature=related http://www.youtube.com/watch?v=lSTg_6N0G7w&feature=related http://www.youtube.com/watch?v=PTlmho_RovY&feature=related*

# INTENTION 8 PROMOTE ECONOMIC EQUALITY

There are over 5 million Canadians living in poverty (15% of the population), of which 1.2 million are children! Around the world, every three seconds a child dies as a result of extreme poverty. The less-developed the country, the larger the human rights violation. Hunger increases as environmental degradation and climate change continue to rise. The UNICEF 2007 report evaluated the welfare of children in the 21 most economically developed countries, and US children came out in the sad position of 20th out of 21 countries.

| ACTIONS | **The Basics** | | | |
|---|---|---|---|---|
| E.8.1 | I will inform myself by downloading a few publications of my interest concerning economic inequality at www.socialjustice.org | | | |
| E.8.2 | I will become more informed about poverty in Canada www.cwp-csp.ca | | | |
| E.8.3 | I will vote for increased social assistance rates. | | | |
| E.8.4 | Every month, 770,000 Canadians use food banks, over 40% of which are children. I will donate ___ items to the local food bank every week for ___ weeks. | | | |

# INTENTION 8  PROMOTE ECONOMIC EQUALITY

| | | PLEDGE DATE | # TIMES/ WEEK | # TIMES COMPLETE / 7 WEEKS |
|---|---|---|---|---|
| ACTIONS | **I Can Do More** | | | |
| E.8.5 | I will advocate for fair taxes. Check out http://www.taxfairness.ca/front for more information and to find out how. | | | |
| E.8.6 | I will sign on with the "Make Poverty History" campaign at www.makepovertyhistory.ca | | | |
| E.8.7 | I will join and support an organization helping the impoverished in Africa. | | | |
| E.8.8 | I will donate on a monthly basis to an organization of my choice helping relieve world hunger. | | | |
| | | | | |
| ACTIONS | **Educating and Influencing Others** | | | |
| E.8.9 | I will advocate for fair and living wages by writing ___letters to the government asking to increase the minimum wage. | | | |
| E.8.10 | I will write ___ letters to government asking them to increase the Child Tax Benefit and ensure that all low-income children receive the benefit of the program. | | | |
| E.8.11 | I will press for trade rules that support human rights and environmental protection. For instance, in my consumer and voting actions I will support trade rules that boost farmers' power in the marketplace. | | | |
| E.8.12 | I will advocate that my government's aid spending is both accountable and transparent. | | | |
| E.8.13 | I will ask the federal government to honour their Millennium Development Goal commitments by setting timetables to reach their foreign aid targets by 2015. See www.endpoverty2015.org | | | |
| | | | | |
| ACTIONS | **Community Projects** | | | |
| E.8.14 | We will organize a fundraiser for a local shelter, foodbank, or community group this month. | | | |
| E.8.15 | We will create a project or group that will advocate for expanded affordable housing and support efforts for adequate social housing. | | | |
| E.8.16 | We will promote the immediate transfer of 100% of interest and debt payments by the poorest countries to be reinvested in these countries on education, health care, and sustainable energy development. For instance, sub-Saharan African countries pay $12 billion per year on interest payments on their loans, whereas that money could be better spent on education, health care and infrastructures. We will look into how we can hold the governments accountable and ensure that the resulting social development is transparent. | | | |

## INTENTION 8  PROMOTE ECONOMIC EQUALITY

| | | PLEDGE DATE | # TIMES/ WEEK | # TIMES COMPLETE / 7 WEEKS |
|---|---|---|---|---|
| E.8.17 | We will research and share information on the advantages of a universally accessible child-care system. | | | |
| E.8.18 | We will get involved with a local initiative to support global change initiatives, specifically: _____ | | | |

## EXPLORATIONS

Download the publications concerning economic inequality at *www.socialjustice.org*

Check out the Poverty and Human Rights Center at *www.povertyandhumanrights.org*

Check out the Canadian government campaign to end poverty by 2000 at *www.campaign2000.ca*

Research the Canadian Council on Social Development at *www.ccsd.ca*

Join the campaign "Dignity for All" for a poverty-free Canada and sign the petition at *www.dignityforall.ca.*

Or find the alternative in your country.

Become more informed about poverty in Canada *www.cwp-csp.ca*

## VALUE E SUPPLEMENT

## INTENTION 3  PARTICIPATORY DEMOCRACY – EXPLORATIONS

Here is an excerpt from Sierra Club BC's "Endangered Species Toolkit " that suggests step by step processes that can kickstart you into political action:

Tools and Tactics is designed to be a quick primer on important skills for citizens working in the public arena. Talking to a public official or dealing with the press for the first time can seem like an ordeal, but you might come to enjoy it sooner than you think! If you have an important meeting or media event coming up, congratulate yourself for doing your homework and getting this opportunity. Take a moment with your group to acknowledge this important step forward. Then take some time to go through the relevant section of this Toolkit together and rehearse potentially tricky situations.

Parts of this section were adapted from former Sierra Club of Canada Executive Director Elizabeth May's handout How to Be an Activist, a distillation of more than twenty years experience campaigning for the environment.

### How to Get Your Message in the News Media

The environmental movement and the news media have an awkward and symbiotic relationship: media need us for stories, and the environmental movement certainly needs the media to reach the public. But many environmentalists are sometimes frustrated by the superficiality of news coverage of issues that threaten planetary survival, and the news media get sick of hearing what they often regard as predictable whining from environmentalists. What can you do to get some reasonably accurate coverage of your concerns? First, you should understand a few things about the news media. As follows are a few suggestions on how to do so.

- Take the reporter's perspective. It helps if you are able to see the story from the reporter's point of view. There are very few newspapers or electronic networks with a full- time environmen tal reporter. You are trying to get a reporter who has to cover everything from tax hikes to crime on the street interested in your story.

## INTENTION 3 **EXPLORATIONS**

- Make the story easy for them. The whirlwind pace of electronic communication, coupled with staff cuts in many - newsrooms, means that there are more stories and less time to write them. An overworked reporter would probably still investigate a corruption scandal, but will give endangered frogs short shrift. If you want a reporter to cover your story, you have to do all their work for them. Think it through. Where's the angle?

- Remember the old adage, "Dog bites man" is not a story. "Man bites dog" is. Make sure your story has an interesting or catchy angle.

- Make sure to fill in the five W's – Who, What, When, Where and Why – for the proposed story. Give your story a "hook".

- Tie your issue to other political events, like elections or previous campaign promises. What are the financial issues? Is taxpayers' money being wasted? Are jobs being lost? Are environmental alternatives better for the economy? (They usually are.) Make it interesting to someone who doesn't fully understand what is at stake.

### Contacting the Media

Below, we've listed four handy routes to take your environmental message to the mainstream media.

### Write Your Own Press Release

Keep in mind that your press release should read like a news story, not like your group's manifesto. Put in quotes from group representatives. Be sure to include phone numbers so that reporters can call you to get more details and rework your press release into their own story. Try to fit it all onto one page – your messages will be more focused and succinct.

Send your release in time to reach the media before or on your release date. If there is cash in your campaign kitty, you can also fax your release to the closest office of Canadian Press (www.cp.org), a wire service. If it puts your story on their service, it will automatically reach television, radio and newspaper newsrooms. It is up to the news director in each outlet to decide whether to use your story.

Beyond press releases, you may want to hold a press conference. This works best if you have a really good story, or if you can bring in an acknowledged expert who wouldn't be available as a matter of course. Hold press conferences somewhere familiar to the media, and make it convenient. Try to avoid having to spend money to rent space. Is there a good community centre close to the downtown? Can you get the help of someone in city council to use City Hall or the Regional Government Centre?

For a sample Press Release, refer to Appendix 2 of the Endangered Species Toolkit.

*"A leader is anyone who wants to help, who is willing to step forward to create change in their world. The leaders we need are already here."* – Margaret Wheatley

# VALUE F
# GREEN MY COMMUNITY

*I value resilient and healthy community and*
*I will help create that for myself and others.*

It is easy to feel overwhelmed and frustrated that what we're doing isn't enough, but we're not alone in our quest to make our world a better place. Starting with our personal actions, supporting each other in Action Circles and on community projects is a fun and empowering way to create systemic change.

As anthropologist Margaret Mead assures us: *"Never doubt that a small group of thoughtful committed citizens can change the world. Indeed, it is the only thing that ever has."*

Working in groups can have its challenges, but the rewards are great. Finding ways to remove the barriers for others to join in is a contagious and invigorating feeling. We feel the strength of community in the creativity and resources that emerge when people gather for common purpose—the result is always greater than the sum of the parts.[18]

It only takes a few to start a movement, but once a certain number of people have been impacted, that way of operating becomes the 'norm' for everyone. This is called a "tipping point," or the "Hundredth Monkey" phenomenon.

Many Action Circles choose to address important local neighborhood issues. Having a common objective, getting to know each other, and working together builds resilience in community. Well-being improves and long-term stability develops. This is an invaluable asset in a world full of change and contributes greatly to a healthy and vibrant way of life.

## I Intend To:

F 1    **Connect-in-Community**

F 2    **Caring Networks and Cultivating a Sense of Place**

F 3    **Co-Create Awareness and Wellbeing**

F 4    **Resilient Local Food Networks**

F 5    **Integrating Nature Back into Community**

F 6    **Vibrant Local Economies**

F 7    **Arts and Celebrations**

F 8    **Community Emergency Preparedness**

---

18. Check out www.soulneeds.com/articles/view/593/The+100th+Monkey for the whole story.

# INTENTION 1 CONNECT-IN-COMMUNITY

This concept is simple: Talk to your neighbours, networks and greater communities and see what happens! When we open up to our communities, we open up ourselves to connection with the world around us. In being in position to support others and receive support and insight, the way you feel and what you can accomplish is astounding. Overall, this means a happier and more fulfilling way of life. The first step to this is connecting with those around you –have fun with this!

| | | PLEDGE DATE | # TIMES/ WEEK | # TIMES COMPLETE / 7 WEEKS |
|---|---|---|---|---|
| ACTIONS | **The Basics** | | | |
| F.1.1 | I will talk to or visit __ people in my community this week. | | | |
| F.1.2 | I will take time this week to consider a local community issue that I would like to get involved in. | | | |
| F.1.3 | I will look into a local community organization, like my local Transition Town Initiative, to get connected with the people in my neighbourhood. 🌍 | | | |
| F.1.4 | I will explore what programs and initiatives my local government offers to help build community, and take advantage of what is offered. 🌍 | | | |
| ACTIONS | **I Can Do More** | | | |
| F.1.5 | I will investigate the benefits of living in shared housing. 🌍 | | | |
| F.1.6 | I will put a bench out on my property where myself and others can sit and connect. | | | |
| F.1.7 | I will consider, talk to my neighbours about, or go ahead with removing my backyard fence so that I can be more connected to the people around me. | | | |
| F.1.8 | I commit to doing action #(s) ____ consistently for ____ days/weeks/months. | | | |
| ACTIONS | **Educating and Influencing Others** | | | |
| F.1.9 | I will discuss the community issue I am most interested in addressing and invite others to join me to meet regularly in an Action Circle that will focus on that particular issue, within our Transition Town Initiative. | | | |
| F.1.10 | I will ask a friend, coworker or family to join me in completing action #(s) _____ from this intention. | | | |
| ACTIONS | **Community Projects** | | | |
| F.1.11 | We will join a Transition Town Initiative Village Building Team and help others start Transition Villages in their neighbourhoods. 🌍 | | | |
| F.1.12 | If a Transition Town Initiative doesn't yet exist in our town or community, we will look into how to start one. 🌍 | | | |
| F.1.13 | If there isn't a Transition "Village" yet in our neighbourhood, we will look into how to start one. 🌍 | | | |

# INTENTION 1 **CONNECT-IN-COMMUNITY**

## EXPLORATIONS 🌐

The global Transition Town Network helps communities deal with climate change, shrinking supplies of energy and resources ("peak everything"), and increasing economic instability and insecurity. This process, which they call Transition, aims to create stronger, happier, more resilient communities. Find out how to join or form a Transition Town in your community: *www.transitionnetwork.org*

Village Vancouver, Vancouver's vibrant and active Transition Town Initiative, connects people together in their neighbourhoods ("villages") and around common interests through potlucks, workshops, events, projects, working groups, networks, and Permaculture. Thousands of people participate in this volunteer run, people powered collaborative effort in building community self-reliance. If you're in Vancouver (or anywhere nearby), you can get involved! *www.villagevancouver.ca*

Information on neighbourhood villages can be found at:
*www.villagevancouver.ca/page/villages-1 and www.villagevancouver.ca/group/vvtransitionvillagebuildinggroup*

SPEC (The Society Promoting Environmental Conservation), Canada's oldest environmental organization, works to create healthy, sustainable communities through education, advocacy, and outreach: *www.spec.bc.ca*

Langara College's Summer School on Building Community, held in June every year, offers a wide variety of classes, workshops, and other activities aimed at building just and sustainable community:
*www.langara.bc.ca/social-cultural/summer-school/index.html*

Collective Houses BC:
*www.villagevancouver.ca/group/collectivehousesbc and www.facebook.com/group.php?gid=10084638013*
Greenest City 2020 (City of Vancouver) *http://vancouver.ca/greenestcity/index.htm*

Hopkins, Rob. *The Transition Handbook: From oil dependency to local resilience.* White River Junction, VT: Chelsea Green Publishing, 2008.
Chiras, Dan and Dave Wann. *Superbia: 31 ways to create sustainable neighborhoods.* Gabriola Island, B.C.: New Society Publishers, 2003
Walljasper, Jay. *The Great Neighbourhood Book: A Do-it-Youself Guide to Placemaking.* Gabriola Island, B.C.: New Society Publishers, 2007.
Engwicht, David. *Street Reclaiming: Creating Livable Streets and Vibrant Communities.* Gabriola Island, B.C.: New Society Publishers, 1999.
Kellogg, Scott and Stacy Pettigrew. *Toolbox for Sustainable City Living: a do-it-ourselves guide.* Cambridge, MA: South End Press, 2008.

In Transition 1.0: from oil dependency to local resiliency (DVD, 2009, 50 min.)
The Power of Community: How Cuba Survived Peak Oil (DVD, 2006, 53 min., Faith Morgan, Dir.)

*"If we do not have peace, it is because we have forgotten that we belong to each other."* – Mother Teresa

# INTENTION 2  CARING NETWORKS, A SENSE OF PLACE

Once we are connected, there is a stronger likelihood that people will look-out for each other, and that means a safer and more caring life experience. We can take this one-step further by instituting projects that build networks that focus on care and attention for the community's well-being overall.

| ACTIONS | The Basics | PLEDGE DATE | # TIMES/ WEEK | # TIMES COMPLETE / 7 WEEKS |
|---------|-----------|-------------|---------------|----------------------------|
| F.2.1 | I will offer to help neighbours when they need assistance with a project. | | | |
| F.2.2 | I will greet new neighbours and welcome them to the neighbourhood. | | | |
| F.2.3 | I will connect with some friends and neighbours to create a Circle focused on creating a 'caring network' in our neighbourhood. | | | |
| F.2.4 | I will think about what we would like to see in a neighbourhood Caring Network (open for a name that suites your group and neighbourhood) and take steps with others to initiate at least one of these ideas. 🌐 | | | |

| ACTIONS | I Can Do More | | | |
|---------|--------------|---|---|---|
| F.2.5 | I will research what a "Mapping Skills" Assessment is. | | | |
| F.2.6 | I will complete, or convene people, to complete a Mapping Skills Assessment of our community. 🌐 | | | |
| F.2.7 | I will watch out for our neighbours' homes/apartments and well-being to make sure they are well, particularly elderly neighbours. | | | |

| ACTIONS | Educating and Influencing Others | | | |
|---------|----------------------------------|---|---|---|
| F.2.8 | I will encourage neighbourhood elders to share their knowledge with others in the neighbourhood. | | | |
| F.2.9 | I will seek out neighbours who have lived in the neighbourhood for a long time and invite them to tell our circle about the history of the neighbourhood. | | | |
| F.2.10 | I will ask a friend, coworker or family to join me in completing action #(s) _____ from this intention | | | |

| ACTIONS | Community Projects | | | |
|---------|-------------------|---|---|---|
| F.2.11 | We will think about what we would like to see in a neighbourhood Caring Network and take steps with others to initiate at least one of these ideas. 🌐 | | | |
| F.2.12 | We will host a "Block Watch Care Community" to engage the community on ways to help each other out. | | | |

# INTENTION 2 **CARING NETWORKS, A SENSE OF PLACE**

| | | PLEDGE DATE | # TIMES / WEEK | # TIMES COMPLETE / 7 WEEKS |
|---|---|---|---|---|
| F.2.13 | We will help form a group where a few or several neighbours help one another on a regular basis. | | | |
| F.2.14 | We will compile a list of elderly neighbours and make a schedule to ensure that they are visited regularly by members of our circle. | | | |
| F.2.15 | We will learn about what streams used to run through our neighbourhood and participate in "lost stream" walks and workshops with others from our community.  | | | |
| F.2.16 | We will explore the work of City Repair in Portland and the Public Space Network in Vancouver, and engage with others in acting to enhance community space and quality of life in our neighbourhood.  | | | |
| F.2.17 | We will personally implement action #(s) _____ from this intention at our schools and/or workplaces. | | | |

## EXPLORATIONS

Village Vancouver Community Asset Mapping Project *http://www.villagevancouver.ca/page/village-vancouver-mapping* False Creek Watershed Society recently completed a community mapping project: *The Water Beneath Our Feet: Mapping the Spirit of the False Creek Watershed.* They also conduct community "lost stream" walking tours and workshops and other place-based educational activities around natural and human history, urban sustainability, and local/1st Nations culture. *www.villagevancouver.ca/group/vancouverwatershedgroup* and *www.falsecreekwatershed.org/* Help explore neighbourhood Caring Networks in Village Vancouver's Heart and Soul Group: *www.villagevancouver.ca/group/vvheartandsoulgroup*

City Repair (Portland) "educates and inspires communities and individuals to creatively transform the places where they live. City Repair facilitates artistic and ecologically-oriented 'placemaking' through projects that honor the interconnection of human communities and the natural world." *http://cityrepair.org/*

Vancouver Public Space Network VPSN is "here to preserve and celebrate public space as an essential part of a vibrant, inclusive city. The VPSN is a grassroots collective that engages in advocacy, outreach and education on public space issues in and around Vancouver, British Columbia." *http://vancouverpublicspace.ca/*

VV Bioregional Network: *www.villagevancouver.ca/group/Bioregional*

Alexander, Christopher and Sara Ishikawa and Murray Silverstein with Max Jacobson, Ingrid Fiksdahl-King, and Shlomo Angel. *A Pattern Language: Towns, Buildings, Construction.* New York, N.Y.: Oxford University Press, 1977.

Kretzmann, John P. and John L. McKnight. *Building Communities From the Inside Out: A Path Toward Finding and Mobilizing a Community's Assets.* Evanston, IL: Center for Urban Affairs and Policy Research, Neighbourhood Innovations Network, Northwestern University, 1993.

Collectively edited. *City Repair's Placemaking Guidebook: Creative Community Building in the Public Right of Way,* 2nd ed. Portland, OR: The City Repair Project, 2006.

*"The good life is inspired by love and guided by knowledge"* – Bertrand Russell

# INTENTION 3 CO-CREATE AWARENESS & WELL-BEING

It is inspiring to share information with others on a day-to-day basis and work together towards common goals. We can even take this benefit further, in our organizations and workplaces. As we enhance and share our awareness about sustainability, we can nurture our own-sense of awareness and personal well-being. The more we embody wellness, the more we reflect that onto the people and the world around us. Based on basic principles of natural law, we know that everything is connected. The inner and the outer world are the same. As we change ourselves, we change the world. These actions can be implemented into a workplace, network or organization in which you are part of or often even your own home environment.

| ACTIONS | The Basics | PLEDGE DATE | # TIMES/ WEEK | # TIMES COMPLETE / 7 WEEKS |
|---|---|---|---|---|
| F.3.1 | I will give full-attention to how much time I spend this week doing feel-good things for myself at my organization. For example: taking a walk on a break, stretching when I have time, or being creative by drawing or doodling at lunch. I will journal on the changes that I would like to make. | | | |
| F.3.2 | I will re-organize, clean, and/or re-fresh my work-space or an area in the building. | | | |
| F.3.3 | I will make ___ enhancement(s) to the décor of my personal space and/or to the building. | | | |
| F.3.4 | I will research and examine the ergonomics at my organization or work-station. 🌍 | | | |

| ACTIONS | I Can Do More | | | |
|---|---|---|---|---|
| F.3.5 | I will make ___ enhancement(s) to the layout of my personal space or to the building. (It's time to get out those Feng Shui books that you got for your birthday!). | | | |
| F.3.6 | I will re-organize physical files and/or storage space. | | | |
| F.3.7 | I will re-organize the folders and delete old messages in my email Inbox. | | | |
| F.3.8 | I will not use my cell phone when it is not needed. | | | |
| F.3.9 | I will bring awareness to the air quality in my work area and/or building to determine if it needs improvement. Is the air stagnant and stuffy, or is it clean and clear. | | | |
| F.3.10 | I will brainstorm the elements at work that make or could make the space productive, happy, healthy and balanced. | | | |
| F.3.11 | I will hold an intention and visualization of the feel and aspects that I want my workplace or organization to have. I will do this ___ times a week/month. | | | |
| F.3.12 | I commit to doing action #(s) ___ consistently for ___ days, weeks or months. | | | |

# INTENTION 3 **CO-CREATE AWARENESS & WELL-BEING**

| | | PLEDGE DATE | # TIMES/ WEEK | # TIMES COMPLETE / 7 WEEKS |
|---|---|---|---|---|
| ACTIONS | **Educating and Influencing Others** | | | |
| F.3.13 | I will promote ___ environmental initiative(s) at a meeting(s), on a message board or in a high-traffic area to others in person and/or via e-mail. | | | |
| F.3.14 | I will schedule a wellness or a stress reduction workshop or I will bring workshop research ideas to a decision-maker. | | | |
| F.3.15 | I will ask a friend, coworker or family to join me in completing action #(s) _____ from this intention. | | | |
| ACTIONS | **Community Projects** | | | |
| F.3.16 | We will research organizational wellness or stress reduction workshops and follow through on organizing one. | | | |
| F.3.17 | We will research ways that our workplaces and organizations can enhance their sustainability principles regarding employee workload, and present our findings to a decision maker.  | | | |
| F.3.18 | We will make any necessary ergonomic adjustments at our organizations or workstations. If needed, we will speak to a decision-maker about the necessary ergonomic adjustments as needed. | | | |
| F.3.19 | We will make ___ enhancement(s) to the lighting quality at our workplaces. (For example, using natural lighting where appropriate or turning off the fluorescents in an office and adding a lamp with an efficient light bulb). | | | |
| F.3.20 | Should it be needed, we will create a plan for enhancing the air quality of our workplaces. For example: opening windows/doors for cross-ventilation, blinds/ curtains/awnings, or buying energy efficient fans, humidifiers, air filtering appliances, or an air quality monitor.  | | | |
| F.3.21 | We will organize and manage environmental health and safety audits at our workplaces. | | | |

# EXPLORATIONS

Air Quality Information – Public Serves and Works – Government of Canada: *http://www.tpsgc-pwgsc.gc.ca/biens-property/qcnsl-qtps/qcnsl-qtps-eng.html*

Sustainability Coordinator Information:

"The Green Workplace" by Leigh Stringer

The Green Workplace: *www.thegreenworkplace.com/*

Office ergonomics – Occupational Health and Safety: *http://ohs.csa.ca/standards/ergonomics/General/dsp_General-Workplace.asp* *www.oecd.org/dataoecd/34/27/42944011.pdf*

Canadian Centre for Occupational Health and Safety: *www.ccohs.ca/*

The University of Toronto's Environmental Health and Safety website: *www.ehs.utoronto.ca*

# INTENTION 4 RESILIENT LOCAL FOOD NETWORKS

As the globalization of our food supply increases, so too does the inherent unnecessary use of resources. It is simply not sustainable to be importing food items that can be produced or grown locally. That's just one of many ways in which local food networks can come into play. Together we can become more self sufficient and reliant, growing together as a community - nurturing each other and ourselves!

| Actions | The Basics | PLEDGE DATE | # TIMES/ WEEK | # TIMES COMPLETE / 7 WEEKS |
|---------|------------|-------------|---------------|----------------------------|
| F.4.1 | I will research the benefits of the two-block diet or the 50-mile diet. 🌐 | | | |
| F.4.2 | I will get together with a few other people and create an Action Circle that is focused on creating food networks. | | | |
| F.4.3 | I will research what the benefits of a local neighbourhood food network are. 🌐 | | | |
| F.4.4 | I will research a local community food issue where we, as a group, can get involved. | | | |
| F.4.5 | I will join a local food network! 🌐 | | | |

| Actions | I Can Do More | | | |
|---------|---------------|--|--|--|
| F.4.6 | I will learn more about the benefits of bee keeping in our community. | | | |
| F.4.7 | I will explore the opportunities for having backyard chickens. 🌐 | | | |
| F.4.8 | I will explore possible collaborations with backyard chickens, such as joint purchasing. | | | |
| F.4.9 | I will learn at least two food preservation methods, including canning. | | | |
| F.4.10 | I will think about and brainstorm what our neighbourhood might look like if it were "food secure". | | | |
| F.4.11 | I will think about and brainstorm what our neighbourhood might look like if it were "food resilient." | | | |
| F.4.12 | I will eat a 50-mile diet ___ times this week. | | | |
| F.4.13 | I will eat a two-block diet ___ times this week. | | | |
| F.4.14 | I/we will help someone turn their lawn into a food producing garden. | | | |

| Actions | Educating and Influencing Others | | | |
|---------|----------------------------------|--|--|--|
| F.4.15 | I will organize a tour de coup to visit neighbourhood backyard chickens. 🌐 | | | |
| F.4.16 | I will create a blog or other means of social media that others can follow in our/ my progress in community involvement and projects. | | | |

# INTENTION 4 **RESILIENT LOCAL FOOD NETWORKS**

| | | PLEDGE DATE | # TIMES/ WEEK | # TIMES COMPLETE / 7 WEEKS |
|---|---|---|---|---|
| F.4.17 | I will find or create website(s) where we can post our Best Practices regarding local food networks and production so that the broader community may learn from our experiences. 🌐 | | | |
| F.4.18 | I will find out ways to support food justice efforts in our neighbourhood and engage in them. 🌐 | | | |
| F.4.19 | I will ask a friend, coworker or family to join me in completing action #(s) _____ from this intention. | | | |
| ACTIONS | **Community Projects** | | | |
| F.4.20 | We will join or help start a neighbourhood-based seed savers collective. 🌐 | | | |
| F.4.21 | We will participate in a series of "garden building" work parties – many hands make light work. 🌐 | | | |
| F.4.22 | We will organize a community canning party. | | | |
| F.4.23 | We will get actively involved with beekeeping in our community. If there is no beekeeping in our community, we will propose the idea. 🌐 | | | |
| F.4.24 | We will start a collaborative garden in our neighbourhoods. 🌐 | | | |
| F.4.25 | We will help expand urban farming and market garden endeavors in our neighbourhoods. 🌐 | | | |
| F.4.26 | We will explore the possibilities of utilizing streets to grow more food in our neighbourhood through-market, community, and collaborative garden. 🌐 | | | |
| F.4.27 | We will help organize and hold a "pocket market" in our neighbourhoods. | | | |
| F.4.28 | We will request that our workplaces and/or schools consider implementing action #(s) _____ from this intention; we will research how this is done to make the process as easy as possible. | | | |

# EXPLORATIONS

In Vancouver, you will find several Village Vancouver Neighbourhood Food Networks: *www.villagevancouver.ca/group/villagevancouverfoodworkinggroup* and *5 Food Justice networks: The Westside Food Security Collaborative, Trout Lake/Cedar Cottage Food Security Network, Grandview/Woodland Food Connection, Renfrew/Collingwood Food Security Institute,* and *The Right To Food Network in the Downtown East Side.*

Vancouver's Food Policy Council has a Neighbourhood Food Network Working Group which helps support the work of existing and emerging networks: *http://www.vancouverfoodpolicycouncil.ca/*

Visit Village Vancouver to share your Best Practices and other stories around chickens, bees, seed saving, and community gardening: Backyard Chickens: *http://www.villagevancouver.ca/group/vvfoodgroupseedsavers*

Tour de Coup Vancouver example:

*www.villagevancouver.ca/group/vvfoodgroupcoopcoops/forum/topics/backyard-chickeners-wanted-for-1*

Bees: *http://www.villagevancouver.ca/group/beekeeping*

Seed Savers: *www.villagevancouver.ca/group/vvfoodgroupseedsavers*

# INTENTION 4 **RESILIENT LOCAL FOOD NETWORKS**

Cedar Cottage Seed Savers Collective:
*www.villagevancouver.ca/group/ccseedsavers* and *http://ccseedsavers.wordpress.com/*
Farm Folk/City Folk BC Seeds Project: *www.ffcf.bc.ca/programs/farm.html*
Environmental Youth Alliance (EYA) Urban Seed Project: *www.eya.ca/urban-seeds.html*
Vancouver Plant and Seed Exchange Network: *vanseedtrade.theforum.name/*
Community Gardening: *www.villagevancouver.ca/group/communitygardening*
Two Block Diet: *http://twoblockdiet.blogspot.com/*
Urban Farming/Market Gardening: *http://www.villagevancouver.ca/group/urbanmarketgardens*

Flores, H.C. *Food Not Lawns: How to Turn Your Yard into a Garden and Your Nneighbourhood into A Community.*
White River Junction, VT: Chelsea Green Publishers, 2006.
Pinkerton, Tamzin and Rob Hopkins. *Local Food: How to Make it Happen in Your Community.* Totnes, Devon, UK:
Transition Books/Green Books, 2009.

## INTENTION 5 BRING NATURE BACK INTO COMMUNITY

Old habits are hard to break. It's time to upgrade the old operating system for a new
life-enriching one that holds sustainable principles at the core. Tropical gardens were
once the ideal, but now we are finding more ways to encourage natural local ecosystem processes into our communities, homes, and gardens.

| Actions | The Basics | PLEDGE DATE | # TIMES/ WEEK | # TIMES COMPLETE / 7 WEEKS |
|---|---|---|---|---|
| F.5.1 | I will learn about edible landscaping. | | | |
| F.5.2 | I will learn about what plants and edible landscapes are indigenous to my area. | | | |
| F.5.3 | I will plant local/indigenous plants and shrubs around our homes. | | | |
| F.5.4 | I will familiarize myself with Permacuture Design concepts, especially Social Permaculture. 🌐 | | | |
| F.5.5 | I will get together with some neighbours and create a circle to focus on integrating nature back into our community. | | | |

| Actions | I Can Do More | | | |
|---|---|---|---|---|
| F.5.5 | I will learn about invasive plants species. | | | |
| F.5.6 | I will attend a Permaculture PET (or "blitz") day. These are educational community "barn-raising" work parties which transform a property using Permaculture design principles. 🌐 | | | |

# INTENTION 5 **BRING NATURE BACK INTO COMMUNITY**

| | | PLEDGE DATE | # TIMES/ WEEK | # TIMES COMPLETE / 7 WEEKS |
|---|---|---|---|---|
| ACTIONS | **Educating and Influencing Others** | | | |
| F.5.7 | I will ask a friend, coworker or family to join me in completing action #(s) _____ from this intention. | | | |
| | | | | |
| ACTIONS | **Community Projects** | | | |
| F.5.8 | We will take a permaculture course and encourage others in our group to pursue a Permaculture Design Certificate. | | | |
| F.5.9 | We will start a neighbourhood initiative that addresses invasive species issues in our community. | | | |
| F.5.10 | We will look into urban forest management and personally get involved with the management of our community green spaces; we will also invite our community to get involved and informed in a collaborative manner. | | | |
| F.5.11 | We will request and be actively involved in our workplaces and/or schools switching to local non-invasive gardening, edible landscaping, permaculture or xeroscaping. | | | |
| F.5.12 | We will organize and develop green rooftop initiatives for any of the buildings we occupy regularly (homes, workplaces, etc.). | | | |
| F.5.13 | We will research and co-ordinate LEED certification for any of the buildings we occupy regularly. | | | |

## EXPLORATIONS

Permaculture:

http://www.meetup.com/The-Vancouver-Permaculture-Meetup-Group/

http://www.villagevancouver.ca/page/vv-permaculture-design

http://www.cityfarmer.info/2009/10/01/permablitz-eating-the-suburbs-one-backyard-at-a-time/

PET day example:

www.villagevancouver.ca/events/kits-transition-village-3

Holmgren, David, *Permaculture: Principles and Pathways Beyond Sustasinability.* Hepburn, Victoria, Australia: Holmgren Design Services, 2002.

Mollison, Bill. *Permaculture: A Designers' Manual.* Tyalgum, Australia: Tagari Publications, 1988.

Green Roofs for Healthy Cities: *www.greenroofs.org/*

Canada Green Building Council – LEED Certification: *www.cagbc.org*

*"Happiness is when what you think, what you say,*

*and what you do are in harmony."* – Mahatma Gandhi

# INTENTION 6 VIBRANT LOCAL ECONOMIES

People all around the world are organizing themselves to reclaim the economy from large profit-driven enterprises into sustainable, local alternatives. Focusing on the assets of the local community minimizes dependence on external factors and calls forth a personal power and community-based reliance that offers stability far beyond the means of fickle global systems.

| | | PLEDGE DATE | # TIMES/ WEEK | # TIMES COMPLETE / 7 WEEKS |
|---|---|---|---|---|
| ACTIONS | **The Basics** | | | |
| F.6.1 | I will find out about any local community currency projects and get involved. 🌐 | | | |
| F.6.2 | I will learn about and support local co-operatives and social enterprises in our community. 🌐 | | | |
| F.6.3 | I will learn about Underground Markets. 🌐 | | | |
| F.6.4 | I will get together with some friends who are interested in learning about and helping create local economies, and create an Action Circle with this as our focus. | | | |
| | | | | |
| ACTIONS | **Educating and Influencing Others** | | | |
| F.6.5 | I will have a conversation with another community member not involved about my findings. | | | |
| F.6.6 | I will ask a friend, coworker or family to join me in completing action #(s) _____ from this intention. | | | |
| | | | | |
| ACTIONS | **Community Projects** | | | |
| F.6.7 | We will start an underground market initiative in our community. | | | |
| F.6.8 | We will offer "skills and sharing exchanges," i.e., teaching our skills to others in our community, such as how to mend clothes, repair bicycles, fix household items, etc. | | | |
| F.6.9 | We will create a "tool sharing" exchange in our community. 🌐 | | | |

# EXPLORATIONS 🌐

Local Economy Network: *www.villagevancouver.ca/group/localeconomynetwork*

Community Currency/Dunbar Dollar:

*www.villagevancouver.ca/page/open-money-the-dunbar-dollar and www.dunbardollar.com/*

NOW BC Co-op (Neighbours Organic Weekly Buying Clubs delivers to approximately 25 neighbourhood depots in Vancouver and surrounding areas. They strive to "build a sustainable local food system by connecting local farms and processors with consumers and building community around sustainable food choices. The primary focus for NOWBC Co-op's on-line market is seasonal, local, organic foods sourced directly from small farms and processors." They also source "additional local organic grocery items from several local independent wholesalers." *www.nowbc.ca*

BC Co-operative Assn.: *www.bcca.coop*

Underground Market: *http://www.villagevancouver.ca/page/vancouver-underground-markets and http://vancouverundergroundmarket.com/*

# INTENTION 6 **COMMUNITY PROJECTS**

The Vancouver Tool Library (VTL) is a "cooperative tool lending library currently being established (as of April 2011) in East Vancouver. We are motivated by a vision of our community empowered by the tools and skills needed to transform their homes and communities into vibrant spaces that reflect a commitment to sustainability. To get there, we are creating a community resource that will reduce the costs of improving and greening the places in which we live, work, and play.

*"The VTL is still in the development stages, so we haven't opened our doors yet. When we do, we will be equipped with a wide variety of tools for home repair, gardening, and bicycle maintenance, which will be loaned to our members free of charge. Once we get our feet on the ground, we're also planning to offer workshops and community events."*
*http://vancouvertoollibrary.com/*

## INTENTION 7 CREATIVE ARTS AND CELEBRATIONS

Celebrating life is essential to appreciating and bringing joy into your life and others. Being creative and appreciating others creativity, is a wonderful way for you to embrace co-creating in many ways in yours and others lives. Having fun may quite possibly the most effective way to create community and bring people together!

| Actions | The Basics | PLEDGE DATE | # TIMES/ WEEK | # TIMES COMPLETE / 7 WEEKS |
|---|---|---|---|---|
| F.7.1 | I will participate or choose __ celebratory community event(s) to attend. | | | |
| F.7.2 | I will invite __ neighbour(s) to a social or neighbourhood gathering. | | | |
| F.7.3 | I will look into local community event listings to see where we can volunteer. 🌐 | | | |
| F.7.4 | I will foster a creative aspect within me. | | | |
| F.7.5 | I will create a Community Arts & Celebration Circle with my friends and neighbours. | | | |
| F.7.6 | I commit to doing action #(s) ____ consistently for ____ days/weeks/months. | | | |

| Actions | Educating and Influencing Others | | | |
|---|---|---|---|---|
| F.7.7 | I will invite friends, family or community members over to engage in a creative expressive project like painting, dancing, crafts, or sewing. | | | |
| F.7.8 | I will choose to celebrate someone who wouldn't expect it. | | | |
| F.7.9 | I will invite others over to envision a community in which we would like to live, a radical new idea, a project you want to go forth with or the world in which you wish to live. We will be open to co-creating these ideas through holding the vision and getting excited about it together. | | | |
| F.7.10 | I commit to doing action #(s) ____ for ____ times a week/month for a ___ month(s). | | | |

# INTENTION 7 **CREATIVE ARTS AND CELEBRATIONS**

| ACTIONS | Community Projects | PLEDGE DATE | # TIMES / WEEK | # TIMES COMPLETE / 7 WEEKS |
|---------|--------------------|-------------|----------------|----------------------------|
| F.7.11 | We will organize a neighbourhood gathering, such as a potluck, movie night, soup 'n share, dinner co-op, or drop-in spaghetti night, and invite others to attend. 🌐 | | | |
| F.7.12 | We will organize or get involved in a community environmental and ecological arts project. 🌐 | | | |
| F.7.13 | We will create a block party that closes streets to traffic and consider making this a seasonal or annual event. | | | |
| F.7.14 | We will investigate the possibilities of opening our streets to multiple uses and community activities. 🌐 | | | |
| F.7.15 | We will organize a celebration at our workplace or for our work communities. | | | |

## EXPLORATIONS

Visit Village for ideas about how to organize drop-in spaghetti dinners, soup 'n shares, dinner co-ops and other fun neighbourhood events! *www.villagevancouver.ca/group/artsandcelebration*

Car-Free Day offers Vancouver residents an excellent assortment of street events with hundreds of volunteer opportunities and advocates for opening streets to multiple uses and activities: *www.carfreevancouver.org*

Community Arts Council of Vancouver offers a place "for local artists to connect, for community members to have a voice and for the seeds of change to take root." Founded in 1946. *www.cacv.ca*

# INTENTION 8  COMMUNITY EMERGENCY PREPAREDNESS

Building community resilience with all the intentions is key to prevention. However it's also important to plan for any circumstance, as our world in constantly shifting and many communities are dealing with unforeseen crises. Coming together to prepare also helps to create an atmosphere of comfort, safety and awareness within our community. While educating ourselves on how to be prepared for a multitude of situations, we are brought together as a community. So often it is those with the least that suffer the most in emergency situations, so it's good to broaden our perspective of community as well.  If something does happen we are more likely to be to mitigate some of the chaos, be able to help others and know who to go to for our specific needs.

| ACTIONS | The Basics | | | |
|---------|-----------|---|---|---|
| F.8.1 | I will create an Emergency Preparedness Circle with my friends and neighbours. | | | |
| F.8.2 | I will research into what information and actions are out there in the way of being prepared for a multitude of natural disasters. | | | |

# INTENTION 8 **COMMUNITY EMERGENCY PREPAREDNESS**

| | | PLEDGE DATE | # TIMES/ WEEK | # TIMES COMPLETE / 7 WEEKS |
|---|---|---|---|---|
| ACTIONS | **Delving Deeper in our Circle** | | DISCUSSED-WHO TO CONTACT FOR FURTHER AID. | TAKEN ACTION ON GIVEN TOPIC |
| F.8.3 | We will discuss in a community/neighbourhood supported group how we might feed ourselves for short or extended periods following an emergency and then encourage everyone to take the necessary steps together, so that our neighbourhood is prepared. | | | |
| F.8.4 | We will discuss how we might ensure that we have enough water for short or extended periods following an emergency and then encourage everyone to take the necessary steps together so, that our neighbourhood is prepared. | | | |
| F.8.5 | We will discuss how we can provide as much care (1st aid) as possible to those who require it and then encourage everyone to take the necessary steps together so that our neighbourhood is prepared. (For example, as a start, knowing who has 1st aid or other medical training, know where 1st aid and other medical supplies are, etc.) | | | |
| F.8.6 | We will discuss how we might communicate with one another and the "outside" world and then encourage everyone to take the necessary steps together so that our neighbourhood is prepared. | | | |
| F.8.7 | We will discuss how we can prepare in other ways in advance for emergencies, taking into account a variety of potential obstacles and hazards that we might face, and then encourage everyone to take the necessary steps together, so that our neighbourhood is prepared. | | | |
| F.8.8 | We will discuss how we and our offspring and their offspring might adapt to longer term emergencies -- for instance, changes brought on by climate change, or by "peak everything" or by economic instability, and then encourage everyone to think about what kinds of changes we may need to make to prepare for this. For instance, greatly lessening our dependence on fossil fuels, or learning new skills that help us to become more self-reliant as a community. 🌍 | | | |

| ACTIONS | **Educating and Influencing Others** | | | |
|---|---|---|---|---|
| F.8.9 | I will ask my friends and family on what they have done or intend to do, to be prepared for a natural disaster. | | | |
| F.8.10 | I will attend an Earthquake Preparedness workshop (offered for free in Vancouver in various neighbourhoods and in different languages) and I will encourage as many neighbours as possible to attend with me. | | | |
| F.8.11 | I will have an educated discussion on the topic of emergency preparedness with a decision maker at work. | | | |

# INTENTION 8 **COMMUNITY EMERGENCY PREPAREDNESS**

| | | PLEDGE DATE | # TIMES/ WEEK | # TIMES COMPLETE / 7 WEEKS |
|---|---|---|---|---|
| Actions | **Community Projects** | | | |
| F.8.12 | If an earthquake preparedness workshop is not offered in our community in the near future, we will request that one be offered or look into what it would take to organize one. (Additional workshops can be offered to groups of 15 or more.) | | | |
| F.8.13 | We will make a schedule of what actions need to be taken by our emergency preparedness group and will take action to ensure they are completed by_____ date. | | | |

## EXPLORATIONS

*www.villagevancouver.ca/group/emergencypreparedness*
*http://vancouver.ca/emerg/NEPP/*
*http://www.getemergencyprepared.com/*
*http://www.wordpress.peakmoment.tv/conversations/?p=418*
*http://www.mvcommunitypreparedness.org/*

*"The creative and imaginative efforts and actions of everyone of us count, and nothing less than the health of the world hangs in the balance. Now is the time for us to get on with and amplify the work of healing ourselves, our societies and our planet. No intention is too small and no effort insignificant. Every step along the way counts. And, as you will see every single one of us counts."* – Jon Kabat-Zinn

# BIBLIOGRAPHY

**Baldwin, Christina**. Calling the Circle: The First and Future Culture. Newberg, Oregon: Swan Raven Co., 1994.

**Berry, Thomas. Swimme, Brian.** Universe Story. San Francisco, California: Harper/SanFrancisco, 1994.

**Brown, Molly Young, and Joanna Macy.** Coming Back To Life: Practices to Reconnect Our Lives, OurWorld. Gabriola Island, BC: New Society Publishers, 1998.

**Butterfly Hill, Julia., Jessica Hurley.** One Makes the Difference: inspiring actions that change our world. New York, NY: HarperSanFrancisco, 2002.

**Cohen, Michael.** Reconnecting with Nature. Corvallis, WA: Ecopress, 1997.

**The Earth Works Group.** 50 Simple Things You Can Do to Save The Earth. Berkeley, CA: EarthworksPress, 1989.

**Edney, Paul, and Noah Lieberman.** Change the World for Ten Bucks. Gabriola Island, BC: New Society Publishers, 2006.

**Gershon, David.** Low Carbon Diet. Woodstock, New York: Empowerment Institute, 2006.

**Global Education Associates Upper Midwest. Ed.** An Amazing Journey! The Universe and Me. St.Paul, Minnesota: 2004.

**Goodman, Joel, and Matt Weinstein.** Playfair. San Luis Obispo, California: Impact Publishers, 1988.

**Greenpeace Canada.** Greenpeace Green Living Guide. Toronto, Ontario: 2007.

**Kleiner, Art, Charlotte Roberts, Richard B. Ross, Peter M. Senge, and Bryan J. Smith.** The Fifth Discipline Fieldbook. London, England: Nicholas Brealey Publishing, 1994.

**Korten, David.** The Great Turning. San Francisco, CA: Berrett-Koehler Publishers, 2006.

**Lipton, Bruce.** Biology of Belief. Mountain of Love Publishers, 2005.

**Lipton, Bruce H.** (2005) The Biology Of Belief: Unleashing The Power Of Consciousness, Matter And Miracles.

**Marx-Hubbard, Barbara.** Conscious Evolution: Awakening Our Social Potential. Novato, California: New World Library, 1998.

**Northwest Earth Institute.** Choices for Sustainable Living. Discovering a Sense of Place ,Voluntary Simplicity, Global Warming: Changing CO2urse, Healthy Children - Healthy Planet. Menu for the Future, Reconnecting with Earth, Sustainable Systems at Work. Portland, Oregon: Copyright, 2009.

**The Pachamama Alliance.** <u>Awakening the Dreamer, Changing the Dream Symposium.</u> Facilitator Training Programs.

**The Pollution Probe Foundation.** <u>The Canadian Green Consumer Guide.</u> Toronto, ON: McClelland & Stewart, 1989.

**Sahtouris, Elisabet. Lovelock, James.** <u>Earthdance: Living Systems in Evolution.</u> Bloomington, IN: iUniverse, 2000.

**Sanguin, Bruce, Darwin.** <u>Divinity & Dance of the Cosmos.</u> British Columbia: Copperhouse, 2007.

**Suzuki, David, McConnell, Amanda.** <u>The Sacred Balance: Rediscovering our Place in Nature.</u> Vancouver, BC: Greystone Books, 2006

**Swimme, Brian.** <u>The Universe is a Green Dragon.</u> Rochester, Vermont: Bear & Company, 1984.

**Thich Nhat Hanh.** <u>The World We Have.</u> Berkeley, California: Parallax Press, 2008.

**Thompson, Jill, Elizabeth Farries and Ana Simeon.** <u>Endangered Species Toolkit: A Citizen's Guide to Protecting Biodiversity in British Columbia.</u> Vancouver, BC: Sierra Club BC, 2007.

**Wilber, Ken.** <u>A Theory of Everything – An Integral Vision for Business, Politics, Science and Spirituality.</u> Boston, Massachusetts: Shambhala Publications, 2000.

*"We are living through one of the most fundamental shifts in history – a change in the actual belief structure of Western society. No economic, political, or military power can compare with the power of a change of mind. By deliberately changing their images of reality, people are changing the world."* – Willis Harman

CPSIA information can be obtained at www.ICGtesting.com
Printed in the USA
LVOW09s2338170414

382170LV00006B/16/P

9 780986 607608